Comments from teachers who have participated in professional development workshops based on *Exploring Energy with TOYS* and other Teaching Science with TOYS materials:

"With TOYS, science really becomes part of everyday experiences and materials."

Mary White—Monmouth High School, Monmouth, Illinois

"It's amazing how many toys are based upon physics and chemistry principles. Learning science concepts with toys is an exciting adventure for children. Their natural interest and curiosity in science combined with their desire to 'play' with toys provides great motivation to learn."

Jeannie Tuschl—Tulip Grove School, Nashville, Tennessee

"My classes have come to love the Teaching Science with TOYS activities. Whenever I say we are going to do one of these, the classes cheer because they enjoy them so much."

Terri Skudlarek—Mound Elementary School, Miamisburg, Ohio

"What a huge change in attitudes from past years. [My students] love science and can't seem to wait for more experiments."

Sherrie Foster—Glendale Elementary School, Cincinnati, Ohio

"I'm always amazed how students learn and absorb so much from the experiments with toys. As a teacher, I find it is just necessary to put the names to the concepts they are discovering."

Staci Sabato—Delshire Elementary School, Cincinnati, Ohio

"One of the many things that makes a lesson like this so successful is that [my students] have all played with [the Operation®] game many times. The idea that they would be asked to figure out how it works was not only exciting, but challenged them on a level in which they were comfortable. The interest level was extremely high and throughout the lesson everyone remained on task."

Michael Weaver—Snowhill Elementary School, Springfield, Ohio

"These [TOYS] classes will change forever the way you look and feel about science and the world around you. You will see and think and do *science* everywhere you go."

Pam Bauser—Moraine Meadows Elementary School, Dayton, Ohio

Exploring Energy with TOYS

Other Teaching Science with TOYS Books by Terrific Science Press

Investigating Solids, Liquids, and Gases with TOYS

Exploring Matter with TOYS

Teaching Physical Science through Children's Literature

Teaching Chemistry with TOYS

Teaching Physics with TOYS

Exploring Energy with TOYS

Complete Lessons for Grades 4–8

Beverley A.P. Taylor

Miami University Hamilton

Terrific Science Press
Miami University Middletown
Middletown, Ohio

Terrific Science Press
Miami University Middletown
4200 East University Blvd.
Middletown, Ohio 45042
cce@muohio.edu
www.terrificscience.org

10   9   8   7   6   5   4   3

Photographs by Jeff Sabo, Applied Technologies, Miami University, Oxford, Ohio

This monograph is intended for use by teachers, chemists, and properly supervised students. Users must follow procedures for the safe handling, use, and disposal of chemicals in accordance with local, state, federal, and institutional requirements. The cautions, warnings, and safety reminders associated with experiments and activities involving the use of chemicals and equipment contained in this publication have been compiled from sources believed to be reliable and to represent the best opinions on the subject as of the date of publication. Federal, state, local, or institutional standards, codes, and regulations should be followed and supersede information found in this monograph or its references. The user should check existing regulations as they are updated. No warranty, guarantee, or representation is made by the authors or by Terrific Science Press as to the correctness or sufficiency of any information herein. Neither the authors nor the publisher assume any responsibility or liability for the use of the information herein, nor can it be assumed that all necessary warnings and precautionary measures are contained in this publication. Other or additional information or measures may be required or desirable because of particular or exceptional conditions or circumstances or because of new or changed legislation.

ISBN 1-883822-33-5

This material is based upon work supported by the National Science Foundation under grant number TPE-9055448. This project was supported, in part, by the National Science Foundation. Any opinions, findings, and conclusions or recommendations expressed in this material are those of the authors and do not necessarily reflect the views of the National Science Foundation.

Contents

Program Staff

Chemistry

Mickey Sarquis
Associate Professor of Chemistry
Director, Center for Chemistry Education
Miami University Middletown
Middletown, Ohio

Jerry Sarquis
Professor of Chemistry
Miami University
Oxford, Ohio

John Williams
Associate Professor of Chemistry
Miami University Hamilton
Hamilton, Ohio

Lynn Hogue
Associate Director
Center for Chemistry Education
Miami University Middletown
Middletown, Ohio

Linda Woodward
TOYS Research Associate
Miami University Middletown
Middletown, Ohio

Physics

Beverley Taylor
Associate Professor of Physics
Miami University Hamilton
Hamilton, Ohio

Dwight Portman
Physics Teacher
Winton Woods High School
Cincinnati, Ohio

Jim Poth
Professor of Physics
Miami University
Oxford, Ohio

Mentors

Cheryl Vajda
Teacher
Stewart Elementary School
Oxford, Ohio

Gary Lovely
Science Teacher
Edgewood Middle School
Hamilton, Ohio

Tom Runyan
Science Teacher
Middletown High School
Middletown, Ohio

Sallie Drabenstott
Teacher
Sacred Heart
Fairfield, Ohio

Terrific Science Press Design and Production Team

Susan Gertz
Document Production Manager

Andrea Nolan
Laboratory Coordinator

Lisa Taylor
Project Management
Technical Writing
Technical Editing
Production

Amy Hudepohl
Technical Editing

Julie Hust
Laboratory Testing

Melora Halaj
Technical Editing
Production

Christine Cowdrey
Production

W. Stephen Heffron
Laboratory Testing

Amy Stander
Technical Editing

Stephanie Beaver
Production

Stephen Gentle
Illustration
Photo Editing
Design/Layout
Production

Jennifer Stencil
Design/Layout
Production

Thomas Nackid
Illustration

University and District Affiliates

Matt Arthur, Ashland University, Ashland, OH
Zexia Barnes, Morehead State University, Morehead, KY
Sue Anne Berger and John Trefny, Colorado School of Mines, Golden, CO
Joanne Bowers, Plainview High School, Plainview, TX
Herb Bryce, Seattle Central Community College, Seattle, WA
David Christensen, The University of Northern Iowa, Cedar Falls, IA
Laura Daly, Texas Christian University, Fort Worth, TX
Mary Beth Dove, Butler Elementary School, Butler, OH
Dianne Epp, East High School, Lincoln, NE
Wendy Fleischman, Alaska Pacific University, Anchorage, AK
Maria Galvez-Martin, Ohio State University–Lima, Lima, OH
Babu George, Sacred Heart University, Fairfield, CT
James Golen, University of Massachusetts, North Dartmouth, MA
Richard Hansgen, Bluffton College, Bluffton, OH
Ann Hoffelder, Cumberland College, Williamsburg, KY
J. Hoyt, Wayland Baptist University, Plainview, TX
Cindy Johnston, Lebanon Valley College of Pennsylvania, Annville, PA
Teresa Kokoski, University of New Mexico, Albuquerque, NM
Karen Levitt, University of Pittsburgh, Pittsburgh, PA
Rusty Meyer, Alaska Pacific University, Anchorage, AK
Donald Murad and Charlene Czerniak, University of Toledo, Toledo, OH
Hasker Nelson, African-American Math and Science Coalition, Cincinnati, OH
Judy Ng, James Madison High School, Vienna, VA
Larry Peck, Texas A & M University, College Station, TX
Carol Stearns, Princeton University, Princeton, NJ
Victoria Swenson, Grand Valley State University, Allendale, MI
Leon Venable, Agnes Scott College, Decatur, GA
Doris Warren, Houston Baptist University, Houston, TX
Richard Willis, Kennebunk High School, Kennebunk, ME
Steven Wright, University of Wisconsin–Stevens Point, Stevens Point, WI

Acknowledgments

The author wishes to thank the Terrific Science Press Design and Production Team and the following individuals who have contributed to the success of the Teaching Science with TOYS program and to the development of the activities in this book.

Reviewers

Donna Conner, Roanoke County Schools, Roanoke, VA
Jerry Meisner, University of North Carolina at Greensboro, Greensboro, NC
Sallie Watkins, University of Southern Colorado (ret.), Pueblo, CO
Linda Woodward, Miami University Middletown, Middletown, OH

Other Program Staff

While all of our program staff contribute to the success of the Teaching Science with TOYS program, we want to specifically acknowledge two who have contributed to this book. Two of the activities in this book, "Drop 'n' Popper" and "Make Your Own Motor," were originally developed by Jim Poth. Linda Woodward served as an internal reviewer for this book.

Teachers

The activities in this and other Teaching Science with TOYS books have been contributed and tested by many teachers in the Teaching Science with TOYS program in addition to those listed as contributors to specific activities. We wish to acknowledge their efforts in making these activities effective and relevant teaching tools. In addition, teachers who have made original contributions to the material in specific activities are listed at the end of those activities.

Foreword

Science is not about reading textbooks and memorizing facts. Science is all about asking questions and trying to find their answers. Students should have the opportunity to formulate questions and carry out investigations. They should practice developing explanations for phenomena based on evidence they have collected. Communication and respect for the ideas of others is also an important part of science. The activities in this book will help you create a learning environment in your classroom in which all these things happen naturally. And, as a bonus, the activities use toys and common household items rather than typical science lab equipment. The toys capitalize on students' natural interest and help them see the connection between the science they do in school and their lives outside of school.

Exploring Energy with TOYS is one of several books to result from the National Science Foundation-funded Teaching Science with TOYS program at Miami University. Teachers from around the country participate in this program to increase their knowledge of physics and chemistry and to enhance their ability to teach hands-on, discovery-based science. Teachers tell us that the use of TOYS materials has had many different impacts on their classrooms. Students now look forward to science time. Attendance is up in general because students don't want to miss something exciting. All students in class are now engaged in science. In fact, many teachers find that students who are not leaders in anything else are often their best science students. Parents are more involved, because students go home and talk about what they did in science class. Students apply science outside the classroom. For example, during a field trip to an art museum, one group of inner city sixth graders was overheard animatedly discussing the center of gravity of a statue.

I and all of our program staff would like to thank all the TOYS teachers for the valuable contributions they have made to the activities in this book by reviewing and classroom testing our ideas. Because of their efforts, you can be sure these curriculum materials weren't just dreamed up by some ivory-tower professor who has never seen the inside of a real classroom. We invite you to discover the joy of teaching science by guiding the investigations of students who will find they just can't wait to do the next experiment.

Beverley Taylor, Co-Director
Teaching Science with TOYS

Exploring Energy with TOYS

Introduction

WHAT IS TEACHING SCIENCE WITH TOYS?

Teaching Science with TOYS is a National Science Foundation-funded project located at Miami University in Ohio. The goal of the project is to enhance teachers' knowledge of physics and chemistry and to encourage activity-based, discovery-oriented science instruction. The TOYS project promotes toys as an ideal mechanism for science instruction because they are an everyday part of students' world and carry a user-friendly message. An important component of the TOYS project is hands-on science workshops in which K–12 teachers gain experience using hands-on activities to teach science at their grade level. Through TOYS workshops, over 900 K–12 teachers nationwide have brought toy-based science into their classrooms, using teacher-tested TOYS activities.

Because not everyone can attend a TOYS workshop, the TOYS project produces written materials, such as this TOYS Teacher Resource Module, so that many more teachers and students can share in the fun and learning of the TOYS project.

BECOMING INVOLVED WITH THE TOYS PROJECT

By using the activities in this book, you are joining a national network of teachers and other science educators who are committed to integrating toy-based science into their curricula. If you have a TOYS Affiliate (a college or university science educator who conducts local TOYS programs) in your area, you may want to get involved in local TOYS programming. (See the list of TOYS Affiliates on page x.)

Another way to extend your TOYS involvement is by attending a Teaching Science with TOYS graduate-credit course for teachers of grades K–12 at Miami University in Ohio. The program uses a workshop-style format in which participating teachers receive instruction in small, grade-level-specific groups. Teachers do not need a science background to benefit from the program; the program faculty review relevant science principles as they are applied to the toy-based activities being featured. Participants spend much of their time exploring hands-on, toy-based science activities.

TOYS courses and programs are open to all science educators of grades K–12. Applicants should be actively teaching or assigned to a science support position. For more information or an application, contact

Teaching Science with TOYS
Miami University Middletown
4200 East University Blvd.
Middletown, Ohio 45042
513/727-3421, cce@muohio.edu, www.terrificscience.org

WHAT ARE TOYS TEACHER RESOURCE MODULES?

TOYS Teacher Resource Modules are collections of TOYS activities grouped around a topic or theme along with supporting science content and pedagogical materials. Each module is prepared for a specific grade range. The modules have been developed especially for teachers who want to use toy-based physical science activities in the classroom but may not have been able to attend a TOYS workshop at the Miami site or one of the Affiliate sites nationwide. The modules do not assume any particular prior knowledge of physical science—complete content review and activity explanations are included.

The topic of this module is mechanical energy and energy conversions. Energy is one of the most abstract concepts in physics and is also one of the most useful, as it ties all the different branches of physics (and other sciences) together. Because of its importance, energy is one of the major content strands identified in the National Science Standards developed by the National Research Council. Many current science curricula include a variety of topics related to energy resource utilization and resulting environmental effects, such as energy conservation, nuclear waste, and global warming. In order to understand the issues involved, students need a basic understanding of energy itself, and the study of mechanical energy is the easiest place to start. This module has been developed for use by teachers at the intermediate and middle school level but can be modified for use by teachers of older students.

HOW IS THIS RESOURCE MODULE ORGANIZED?

This module is organized into three main sections: Content Review, Pedagogical Strategies, and Module Activities. We suggest that you read the Pedagogical Strategies section first to get an overview of the unit that you will be teaching, skim the Content Review section to refresh your memory of the concepts, then read the Module Activity section to understand the activities in detail. After that, we recommend that you reread the Content Review section more closely. The following paragraphs provide a brief overview of these sections.

Pedagogical Strategies
The Pedagogical Strategies section is intended to provide ideas for effectively teaching a unit on mechanical energy. It suggests ways to incorporate the toy-based activities presented in the module into a series of lessons using a constructivist approach and employing a learning cycle philosophy.

Content Review

The Content Review section is intended to provide you, the teacher, with an introduction to (or a review of) mechanical energy concepts. The material in this section (and in the individual activity explanations) is designed to provide you with information at a level beyond what you will present to your students. You can then evaluate how to adjust the content presentation for your own students.

The Content Review section in this module covers the following topics:
- Conservation of Energy
- Energy of Motion
- Energy That Is Stored
- Work: A Way to Transfer Energy
- Other Forms of Energy
- The Energy Crisis

Module Activities

Each module activity provides complete instructions for conducting the activity in your classroom. These activities have been classroom-tested by teachers like yourself and have been demonstrated to be practical, safe, and effective in the typical intermediate or middle school classroom. The first page of each activity provides a photograph of the toy or activity setup. Each activity contains the following information:

• Concept Introduction/ Concept Application/ Synthesis:	The learning cycle stage recommended for this activity in Pedagogical Strategies is indicated, along with a suggested grade-level range.
• Time Required:	An estimated time for conducting the Procedure is listed. This time estimate is based on feedback from classroom testing, but your time may vary depending on your classroom and teaching style. Setup time does not include gathering materials.
• Key Science Topics:	Science topics that are significant to the activity are listed in alphabetical order.
• Student Background:	Background knowledge needed by students prior to doing the activity is listed.
• National Science Education Standards:	This section describes how the activity addresses certain Science as Inquiry and Physical Science Standards.

- Additional Process Skills: The activity provides opportunities for students to use these science process skills, in addition to those listed under National Science Education Standards.

- Materials: Materials needed are listed for each part of the activity, divided into amounts per class, per group, and per student.

- Safety and Disposal: Special safety and/or disposal procedures are listed if required.

- Getting Ready: Information is provided in Getting Ready when preparation (other than gathering materials) is needed before beginning the activity with the students.

- Procedure: The steps in the Procedure are directed toward you, the teacher, and may include more than one option for carrying out the activity. In some cases, possible answers to questions are provided in the steps and are italicized.

- Class Discussion: Guidelines for class discussion are provided when an activity requires detailed discussion for closure.

- Variations and Extensions: Variations are alternative methods for doing the Procedure. Extensions are methods for furthering student understanding of topics.

- Explanation: The Explanation is written to you, the teacher, and is intended to be modified for students. The Explanation supplements the Content Review portion of this book and provides information specific to the activity.

- Assessment: Some activities contain strategies for assessing how well students understand either the concepts presented in the activity or the process skills emphasized in the activity. (Additional assessment information can be found in Pedagogical Strategies.)

- Cross-Curricular Integration: Cross-Curricular Integration provides suggestions for integrating the science activity with other areas of the curriculum. These are primarily intended for self-contained classrooms of grades 4–6.

- References: References, if any, used to write the activity are listed.

- Further Reading: Books that teachers and students can read to extend their understanding of the concepts presented in the activity are suggested.

- Contributors: Individuals, primarily Teaching Science with TOYS graduates, who contributed specific ideas to the activity are listed.

- Handout Masters: Masters for data sheets, observation sheets, and other handouts are provided for some activities.

Notes and safety cautions are included in activities as needed and are indicated by the following icons and type style:

Notes are preceded by an arrow and appear in italics.

Cautions are preceded by an exclamation point and appear in italics.

Many activities contain sample data and/or graphs. This is actual data taken with the toys, not an idealization of what you should theoretically obtain. This sample data is meant to give you an idea of what kind of results to expect. You should not expect your results to exactly duplicate ours.

Annotated List of Energy Activities

Below is a listing and brief summary of the toy-based activities that are included in the Activities section of the module. They are listed in the order in which they are referred to in the Pedagogical Strategies section, which is also the order in which we recommend they be done; however, no single teacher is likely to do all of the application activities. See Pedagogical Strategies for a detailed discussion of how to present these activities. Most of the activities contain ideas for extensions into other parts of the curriculum such as math, social studies, and language arts.

1. **What Makes It Go?** This activity introduces the ideas of work, kinetic energy, and potential energy. Students examine the inside of the Push-n-Go® toys and gain experience with energy concepts by explaining how the toys work. Students practice process skills while investigating variables that affect the distance the Push-n-Go travels.

2. **The Toy That Returns.** Students try to determine how a commercial roll-back works. Then they use coffee cans, rubber bands, and weights to make their own version of this toy. In this toy, kinetic energy is stored temporarily as elastic potential energy in a rubber band and converted back to kinetic energy as the rubber band unwinds. This becomes a model later on for understanding gravitational potential energy by looking at a ball tossed in the air. In this case, kinetic energy is stored as gravitational potential energy and then returns to kinetic energy as the ball comes back down.

3. **How Much Energy?** The toy used in this activity is similar to the Push-n-Go toy used earlier. However, this toy uses a different kind of spring so that more energy can be stored by pushing on the rider several times. Students investigate the relationship between the number of pushes and the distance the toy travels. This activity builds skill in interpreting graphs. Unlike the Explorer Gun® activity (described next), this one produces a linear relationship between number of winds and distance traveled.

4. **Exploring Energy with an Explorer Gun.** Students work in cooperative groups to measure how far the disk "ammunition" of an Explorer Gun goes for different amounts of input energy (number of winds). Students also discover that one wind at the beginning does not store the same amount of energy as one wind near the end, and they connect this to the relationship between work and force.

5. **Pop Can Speedster.** Students make a simple homemade toy that illustrates the conversion of elastic potential energy to kinetic energy.

After making the toy, the students observe its behavior, predict what will happen if the toy is used in different ways, and use their knowledge of energy to describe how the toy works.

6. **Ladybug, Ladybug, Roll Away.** The ladybug in this activity is another homemade energy toy. Both the physics of how the toy works and the manner in which the lesson is carried out are similar to the Pop Can Speedster, so you probably would not choose to do both. This toy is slightly more complicated to make but is easily connected to other areas of study. The toys can be decorated as a variety of animals, such as turtles, grasshoppers, or chickens, to go with a story the class is reading, or as snowmen or pumpkins to help celebrate the season.

7. **Rubber Band Airplane.** This activity is another investigation into the relationship between the amount of stored energy and the distance the toy is able to travel. The toy in this case is a homemade wingless airplane that travels along a track made of fishing line. Since the toy can be made of mostly simple materials (such as straws and rubber bands), the activity requires less investment in materials than the other activities on this topic. It also capitalizes on the normal student interest in anything involving flight.

8. **Slingshot Physics.** This activity uses water balloon slingshots, which are sold at most stores during the spring and summer. A scale made of straws and tape is added to the slingshot so that the distance it is pulled back can be measured. Students then shoot lightweight balls using different amounts of "pullback." Students graph their data and use the graphs to draw conclusions about how far the ball might have gone if they had used still different amounts of "pullback."

9. **The Catapult Gun.** In this activity, the amount of stored energy is constant, and the mass of the projectile is varied. Since the amount of stored energy is always the same, so is the amount of kinetic energy given to the projectile. Kinetic energy depends on both the mass and velocity of the object. For the same amount of kinetic energy, objects with greater mass will have less velocity and so will travel a shorter distance before landing. Process skills of predicting, collecting data, and drawing conclusions from data are used in this activity. One aspect of drawing conclusions from data is considering explanations for anomalous data. Students will have at least one data point that does not agree with the rest and must analyze why this is so.

10. **Loop-the-Loop Challenge.** Several manufacturers of toy cars now sell loop-the-loop tracks, which provide an excellent medium for introducing gravitational potential energy as students explore ways to make the car travel the entire track. Working with these tracks helps students apply their knowledge of how real roller coasters are made to the classroom energy discussion. The activity can also be used to review centripetal force if students have previously studied it.

11. **Homemade Roller Coaster.** This activity extends the study of gravitational potential energy begun in "Loop-the-Loop Challenge." Students initially build tracks with pipe insulation and investigate how the height of the starting point affects how far a marble rolls on the track. After this quantitative part of the activity, students investigate the effects of curves, loops, and hills qualitatively. As a culmination, students design a roller coaster-like track that the marble can successfully negotiate.

12. **Bounceability.** This activity, in which a variety of balls are dropped on different surfaces, illustrates the conversion of gravitational potential energy to kinetic energy and vice versa. More importantly, it initiates thought about what happens to mechanical energy that seems to disappear. Control of variables is emphasized as students design their own experiments. Making and interpreting graphs are also important parts of this activity.

13. **The Energy Transformation Game.** Students use an adaptation of the popular "Guess Who?®" game to think about the energy transformations made by common devices. For example, an iron transforms electrical energy into heat, or thermal energy. Pictures of the devices are shown on the game cards (provided). Teams try to guess their opponent's device by asking questions such as "Does it produce light energy?" and eliminating possibilities. The level of the game may be varied by not using all the possible cards. Many additional ideas are provided for ways to use this game with other subjects.

14. **Drop 'n' Popper.** This activity presents the students with a discrepant event. Having done "Bounceability," they are aware that a ball should never bounce back to its initial height because some of its energy is converted into thermal energy during the collision process. However, the Drop 'n' Popper will bounce back to a point much higher than its initial height. Students must figure out why this occurs. Then they investigate how the surface the Popper is dropped on affects its rebound height. Commercial versions of the Drop 'n' Popper have been sold under a variety of names—most recently as part of a Snailiens™ Battle Set. However, instructions are provided for making them out of racquetballs.

15. **Apply Your Energy Knowledge.** This activity is a follow-up to the exploration phase of the unit. Students are asked to choose a couple of toys from the group available in the learning center and answer a series of questions. These questions guide the students through thinking about different kinds of energy and the energy transformations that take place as one plays with the toys. The activity could also be done by small groups of students working simultaneously on different toys and then sharing their discoveries with the class. It could also be used as a form of assessment at the end of the energy unit.

16. **Doc Shock.** In this application activity, students take apart an Operation® game and investigate how the game transforms electrical energy from a battery into light and sound energy as the game is played. It can also be used to discuss complete circuits, conductors, and insulators.

17. **Make Your Own Motor.** Students make a simple electrical motor that transforms chemical energy into electrical energy and then into mechanical energy. Parallels can be drawn with many motor-driven appliances such as fans and washing machines. If students have studied magnetism, they can investigate how the magnetic force produces the rotation of the coil.

18. **Chemical Energy Transformations.** This activity enables students to observe the transformation of chemical potential energy to several different kinds of energy, such as heat, light, and sound. These conversions are demonstrated by the teacher using common materials and toys such as a lightstick (light), Blaster Balls (sound), a hand warmer or HeaterMeal® (heat), and a candle (light, heat).

19. **Simple Machines with LEGO.** Groups of students invent and build simple devices containing levers, pulleys, and gears. Then groups compete to see who can build (and explain) the device with the most simple machines incorporated into it. Discussion focuses on the fact that simple machines do not let us perform tasks by doing less work or using less energy. Fundamental to their operation is the principle that input and output energy are equal.

20. **Get It in Gear with a LEGO Vehicle.** Students investigate the function of gears by building and discussing two LEGO® vehicles. Students count the teeth on various gears and discover the importance of gear ratios in changing either the speed at which the gear turns or the force it can apply to something else. At the end, students are challenged to change the design of one of the vehicles in some way that will affect how it works.

21. **Squish 'em, Squash 'em, Squoosh 'em.** Students will apply their knowledge of simple machines to analyze how the game Grape Escape® works. This activity can be done as a learning center so that only one game is necessary, or it can be done in small groups.

Safety

Hands-on activities and demonstrations add fun and excitement to science education at any level. However, even the simplest activity can become dangerous when proper safety precautions are ignored, when done incorrectly, or when performed by students without proper supervision. The activities in this book have been extensively reviewed by hundreds of classroom teachers of grades K–12 and by university scientists. We have done all we can to assure the safety of the activities. It is up to you to assure their safe execution!

BE CAREFUL—AND HAVE FUN!

- Activities should be undertaken only at the recommended grade levels and only with adult supervision.

- Always practice activities yourself before performing them with your class. This is the only way to become thoroughly familiar with an activity, and familiarity will help prevent potentially hazardous (or merely embarrassing) mishaps. In addition, you may find variations that will make the activity more meaningful to your students.

- Never attempt an activity if you are unfamiliar or uncomfortable with the procedures or materials involved. If necessary, consult a high school or college science teacher for advice. They are often delighted to help.

- Read each activity carefully and observe all safety precautions and disposal procedures.

- You, your assistants, and any students observing at close range must wear safety goggles if indicated in the activity and at any other time you deem necessary.

- Special safety instructions are not given for everyday classroom materials being used in a typical manner. Use common sense when working with hot, sharp, or breakable objects, such as flames, scissors, or glassware.

- Recycling/reuse instructions are not given for everyday materials. We encourage you to reuse and recycle the materials according to local recycling procedures.

Pedagogical Strategies

This section is intended to provide ideas for effectively teaching a unit on mechanical energy. It suggests ways to incorporate the toy-based demonstrations and hands-on activities presented in the module into a series of lessons based on a learning cycle philosophy. No attempt has been made to indicate what one should do on each day, because the length of a class period varies with the school and with the age of the students. In general, several weeks of activities will be required for the students to master these energy concepts.

THE LEARNING CYCLE

The National Science Education Standards developed by the National Research Council envision science instruction in which students are able to engage in extended investigations. "Science as inquiry" describes both what is taught as well as how it is taught, and teachers orchestrate discourse among students about scientific ideas. This section will help you create a classroom environment in which students can experience science in a way that embraces all these ideals.

The activities in this module follow the Learning Cycle model introduced by the Science Curriculum Improvement Study (SCIS) and since adopted by many other science educators. The stages of the learning cycle have gone by many names over the years. We have chosen to label them Exploration, Concept Introduction, Concept Application, and Synthesis. The upper right-hand corner of the first page of each activity contains a list of the Learning Cycle stages used in this book (except for Exploration). The stage in which the activity is used is indicated by a check mark and a grade range for which the activity is appropriate. Exploration occurs once at the very beginning of the unit and so is not included in the list.

The stages of the Learning Cycle model used in this module are as follows:

Exploration. In this phase, students begin their study of the topic by spontaneously handling and experimenting with objects to see what they do and what happens to them under conditions chosen by the students. This phase is relatively undirected by the teacher in the sense that students choose what they want to do with the objects. However, the teacher can facilitate the experience by asking questions and making comments that encourage the students to explore further. In this module, the Exploration phase comes once, at the beginning of the unit, with the teacher introducing toys and students experimenting with them.

Concept Introduction. In this phase, which follows the Exploration phase, students are placed in situations in which a need for a new concept develops, and they explore the concept without worrying about the formal scientific language that describes it. Afterward, the teacher helps the students solidify their understanding by introducing concepts by name and helping students make connections to their observations and experiences.

Concept Application. In this phase, students apply the concepts they have learned to new situations, which helps them to strengthen their understanding of the concepts and enlarge their meaning. Practice and repeated use in differing situations are required for mastery of a new concept.

Synthesis. This stage gives students the opportunity to integrate several of the concepts they have learned, enabling them to develop a broader understanding of how all of the concepts they have studied fit together. Although this phase is sometimes considered to be part of the application phase, we feel it is sufficiently important to have a category of its own. While we have labeled only one activity at the end of the unit as Synthesis, many of the Application activities involve application for the most recently introduced concept and synthesis with prior ones.

In our zeal to implement hands-on science, we sometimes forget that reading is also an appropriate strategy for learning about science. Although no specific mention of reading has been made in this section, students certainly should do some reading about the various energy topics after the topics have been introduced via the hands-on activities. You may want your class to read materials that supplement topics presented in the Application section. If you have a science textbook which covers energy students can read from it, but don't limit them to the textbook. Most of the activities suggest books that are appropriate for students and can probably be found in school or public libraries. You might want to ask your librarian to assemble a collection of books that include energy topics that your students can browse in their free time. Near the end of the unit you might want to assign students specific topics such as solar or geothermal energy to read about and report back to the class.

INITIATING A NEW TOPIC OF INVESTIGATION

The Push-n-Go toy by Tomy provides an excellent introduction to the ideas of work, potential energy and kinetic energy. You might want to use a dialogue something like this, demonstrating the toy several times as you talk:

> "I brought this toy to class today to help us get started thinking about a new topic. This toy is called a Push-n-Go, which is a sensible name, because I push and it goes. Hmm, that's strange. We

have learned that when I push an object that is sitting still, it always moves in the direction of the force. (Demonstrate by pushing on a book or free-wheeling car.) Yet, in this case, I pushed down and the locomotive moved horizontally across the table. Why do you think that happened?"

You will probably get a variety of answers. Eventually someone's response will give you an opening to say something like, "Oh, so there was a time delay. I pushed, and then sometime later the truck moved, so my force is not what made the truck move. It's like something inside stored up my push and then gave it back later. What do you think might be inside the truck?" You might list the answers on the board. The list will certainly include springs and may include gears depending on the previous experience of the students.

Make a transition to the exploration phase of the lesson by saying something like, "These are interesting ideas. We'll talk more about the Push-n-Go later. First, can you think of other toys or objects that exhibit this same kind of behavior?" Some possibilities are a rubber band shooting a paper wad, pull-back cars such as Hot Wheels® Power Command™ Racers, and spring-up toys (a spring is compressed and held by a suction cup, and later the cup releases and the toy jumps up into the air). The common feature here is that a force is exerted in one direction, then sometime later the object moves in a different direction. Tell the students, "You will be investigating a variety of toys that behave in different ways and looking for things they have in common."

For the exploration phase of the lesson, you will need to assemble a collection of toys whose operation involves stored energy. In addition to the toys mentioned in the previous paragraph, possibilities include a ball, a yo-yo, a bungee cord, a paddleball, any wind-up toy, or any battery-powered toy. A further listing of toys that could be used and a discussion of how each works can be found in the activity, "Apply Your Energy Knowledge."

 This activity itself is not appropriate here. While similar in some ways to the exploration described below, it requires an understanding of energy terms that students have not yet developed.

Students, either individually or in small groups, should play with each toy and hypothesize about how it works, looking for things the toys have in common. For instance, students might group all toys that use rubber bands, all toys that use springs, and all toys that use batteries. In some cases this will just be a guess, because the spring or rubber band is not visible. After all the students have done this, they should be given the opportunity to share their ideas. If you would like the students to make some measurements at this stage, they could measure how far a toy moves after being set in motion or how long the spring-up toy waits before it moves.

WORK, KINETIC ENERGY, AND POTENTIAL ENERGY

Concept Introduction

At this point, the Push-n-Go toys should be disassembled. Depending on the age of the students, they may do it, or you may disassemble them in advance. If you disassemble them in advance, you should keep one together and let the students watch you open it up before distributing the ones that are already open. Have the students investigate how the toy operates. Through discussion bring the students to a consensus about how the toy is able to move in a direction other than the one in which you pushed and what prevents it from moving while you are pushing. Next, discuss the operation of the toy in terms of work, potential energy, and kinetic energy. (Details relative to both of these discussions are provided in the activity, "What Makes It Go?") Be sure to make the point that the greater the speed of an object, the greater its kinetic energy, because this will be important as you proceed. If you think your students will be more comfortable with the terms "stored energy" and "motion energy," then use them. The words are less important than the ideas.

Now the students should complete the rest of the "What Makes It Go?" activity, in which they investigate the dependence of kinetic energy on the mass and speed of the object. In addition, they discover that the toy's kinetic energy seems to disappear as it slows down and stops. Later activities will show how kinetic energy is converted into other types of energy such as thermal energy.

Return to the list the students brainstormed earlier about what might be inside the Push-n-Go toy. Ask them to identify things on the list that they think could store energy. Also, allow students to add other things that might store energy (for example, a trampoline or spiral phone cord). Demonstrate a toy that uses a visible rubber band to store energy. Describe for the students the energy processes taking place as the toy moves. Demonstrate either a commercial or homemade roll-back toy (which contains a hidden rubber band) for the students. Tell the students that they will be allowed to make one of these toys and then will be asked to explain how the new energy concepts they have learned apply to this toy. Complete instructions for making this toy can be found in the activity "The Toy That Returns." Students should be able to articulate that the person using the toy does work by pushing the toy, giving it kinetic energy. As the toy rolls forward, the kinetic energy is gradually stored as elastic potential energy in the tightly wound rubber band. This stored energy is eventually converted back to kinetic energy.

Now the students are ready to investigate what affects the amount of energy stored. It is recommended that your students complete the activities "How Much Energy?" and "Exploring Energy with an Explorer Gun." In "How Much Energy?" toys similar to the Push-n-Go but with variable energy input are used. The students discover that the amount the spring is wound is linearly related to the amount of energy stored (as indicated by the distance traveled). In the Explorer Gun activity, a similar investigation is conducted with different results. The number of winds of the toy is not simply related to the amount of energy stored. It is quite clear that one wind near the end of the winding stores more energy than one wind at the beginning. Students are able to connect this to their kinesthetic experience that the disk gets harder to turn the more they wind it, so they must exert a larger force and thus do more work. This experiment also provides an excellent opportunity to discuss control of variables, variability in measurements, why we average repeated measurements of the same quantity, and the reliability of the results. These ideas are all discussed in the Explanation section of the activity.

Concept Application

At this point in the unit, students should complete one or more hands-on activities that allow them to apply the concepts of work, kinetic energy, and potential energy to new situations. These activities also give students experience with science process skills such as observing, predicting, designing experiments, measuring, graphing, and drawing conclusions from data.

Two additional activities in which students make and investigate the behavior of homemade rolling toys are "Pop Can Speedster" and "Ladybug, Ladybug, Roll Away." Three additional activities in which students can investigate energy storage in a quantitative manner are "Rubber Band Airplane," "Slingshot Physics," and "The Catapult Gun." It is unlikely that any teacher would want to take time to do them all, but several have been provided so that you can choose the ones that seem most interesting to you and your students.

GRAVITATIONAL POTENTIAL ENERGY
Concept Introduction

Once the students are comfortable with the idea that energy can be stored, it is time to introduce different kinds of potential energy. The stored energy in springs that students have already encountered is called *elastic* potential energy. Now you should introduce the students to *gravitational* potential energy. This can be done in a way that students find memorable using the activity "Loop-the-Loop Challenge." After being introduced to the idea that lifting an object can store energy in the object, use the discussion described in the next few paragraphs to further explore the concept with the students. This is probably best done on the day following the "Loop-the-Loop Challenge" activity.

Toss a ball up in the air. Ask the students what kind of energy the ball has. *(Kinetic energy.)* Ask someone to describe the motion of the ball. Be sure the fact that the ball slows down as it goes up then speeds up again as it comes back down comes out in the discussion. Ask the students if the kinetic energy of the ball is constant and, if not, when it has the most kinetic energy. *(It has the most kinetic energy when it is moving with the greatest speed. This is just after it leaves your hand and again just before you catch it.)*

Ask the students how the ball can lose kinetic energy and get it back again. Lead them to make a comparison to the roll-back toy that they built. (If you did not do that activity, you could talk about what happens when someone lands on a trampoline and bounces back up.) Explain that in this case, even though there is no spring or rubber band to store the energy, we can still think of what is happening in terms of stored energy.

Hold the ball out and drop it. Remind the students that because of its position above the ground, the ball has energy that is turned into kinetic energy as it falls and that we call this kind of stored energy gravitational potential energy. The amount of energy the ball has is determined by its weight and how far above the ground it is. Someone may ask why it is called "gravitational" potential energy. We give it this name because the ball would not have this energy in the absence of gravity and would not move when released. In fact, the amount of energy the ball has when held above the ground would be different on Mars or Venus than on the Earth because the size of the gravitational force on the ball would be different.

Concept Application

The activity "Loop-the-Loop Challenge" will probably have stimulated students' interest in how roller coasters work. After doing this activity, students should be allowed to work on the activity "Homemade Roller Coaster." In this activity they will explore how the height of the first hill affects the amount of energy stored and the subsequent speed of the marble. In addition, they will look at what factors affect the ability of the marble to complete a loop or barrel roll. Friction will again enter the discussion as something that is important to the apparent loss of energy of the marble as it moves through the track. This idea sets the stage for the following discussion of other forms of energy and energy conservation.

OTHER FORMS OF ENERGY, ENERGY TRANSFORMATIONS, AND ENERGY CONSERVATION

Concept Introduction

Ask the students to make a list of the cases they have seen during your study of energy in which an object lost its kinetic energy and that energy did not appear to be stored anywhere. After they have done that, ask them to add to the list similar events they can think of in everyday life. (One example could be when a car stops at a red light.) Ask the students to try to think of something all these events have in common. If the students do not come up with the idea that all cases involve friction between the object that has kinetic energy and something else such as the ground, suggest it yourself. Have students rub their hands together and note that their hands feel warm. Friction causes kinetic energy to be turned into another form of energy that is commonly called heat. If you did not do it at the time, you may want to get the Push-n-Go toys back out and let pairs of students send the toy rapidly back and forth between them and notice that the wheel rings get warm. With younger students, you might remind them of their experiences with clay getting warm when rolled on a table. Heat (more properly called thermal energy) is not a form of stored energy because it cannot easily be changed back into kinetic energy.

Ideally, at this point you would do experiments in which the amount of kinetic energy and the amount of heat produced were measured and shown to be equal. Unfortunately, these types of measurements are very difficult to do. Therefore, you should at this point just tell the students that scientists have shown many times that when energy is converted from one form to another, no energy is ever lost. The total amount of energy always remains the same no matter how many times it changes form. Students should now complete the activity "Bounceability." This activity gives students a chance to practice observing energy transformations as well as designing experiments and interpreting data.

Before moving to the next activity, you should introduce the students to other forms of energy. Ask the students to give some examples of other toys in which energy is changed from one form to another. You may want to extend the students' thinking to non-mechanical kinds of energy, such as the chemical energy stored in a battery and later turned into electrical energy, or the chemical energy stored in the food you eat, which then allows your muscles to move and do work. When this discussion slows down, ask students if they can think of anything else that might be a form of energy. Many of them will have heard of solar energy, for instance. You may want to break this down for them into visible light and the other forms of electromagnetic waves we receive from the sun, such as ultraviolet. The energy carried by these non-visible waves is often called radiant energy. Someone may suggest sound as a form of energy, and this response is acceptable since sound is a wave and waves do carry energy. While we often talk about heat energy and sound energy as separate kinds of energy, both can also be described as kinetic energy of moving molecules. To illustrate another form of energy students have probably not considered, hold a paper clip or nail near a magnet, then release it. The students should observe that it moves and so must have kinetic energy. Then they should think about where that energy came from. We call the energy that the paper clip had because of its closeness to a magnet "magnetic potential energy." This energy is analogous to the gravitational potential energy that the ball had because it was near the Earth, which exerts a gravitational force on it.

Students should now play "The Energy Transformation Game." Be sure you have discussed all the forms of energy that are included on the cards you have chosen to use.

At this point, you may want to introduce issues of the "energy crisis" the world faces because we are "running out of energy." If students have encountered this idea in previous studies or on television, they may be confused by your assertion that energy is never lost. You need to help them understand that the important issue in the "energy crisis" is that energy that is in forms that are easy for us to use, such as the chemical energy in natural gas and coal, are rapidly being transformed into forms such as heat that cannot readily be reused. The book *What Makes Everything Go?* by M.E. Ross can be very helpful in this discussion and is short enough to read aloud in class. You may want to have students do some reading on ways in which electricity is produced; alternative forms of energy, such as solar and wind energy; and patterns of change in world population and energy usage. Each student should then report back to the class on what he or she learned.

Concept Application

Several additional activities are included to extend student thinking about energy transformations. "Drop 'n' Popper" is a fun extension to the "Bounceability" activity. "Apply Your Energy Knowledge" can be done in cooperative groups or individually as students have free time. Most of the attention of this unit focuses on mechanical energy, because it is the least abstract form of energy and thus is more appropriate for the introduction of basic concepts. If you would like to give students more experience with other important forms of energy, three activities are available that deal with electrical and chemical energy: "Doc Shock," "Make Your Own Motor," and "Chemical Energy Transformations."

You may want to do some activities involving simple machines at this point. It is not essential that simple machines be included as part of a unit on mechanical energy; however, they often are because their basic function is to transfer energy. If your curriculum already includes simple machines, or if the students have previously studied simple machines, three activities can provide a good bridge between the two topics: "Simple Machines with LEGO," "Get It in Gear with a LEGO Vehicle," and "Squish 'em, Squash 'em, Squoosh 'em."

CONCLUSION AND ASSESSMENT

To conclude this unit, students should be given the chance to pull together everything they have learned. The "Apply Your Energy Knowledge" activity provides one way to do this. A less structured alternative is to provide the students with a wide range of toys and ask them to identify the energy transformations that take place in each toy. For example, a cap gun changes chemical energy into heat and sound energy, and a wind-up walking toy changes elastic potential energy into kinetic energy. This could be done in a cooperative group situation, with each student in the group investigating a different toy. Each student should eventually write a description of the energy transformations in his or her toy, but everyone should have the opportunity to discuss the toy with his or her group first.

Additional ideas for demonstrations involving energy and energy transformations can be found in many books on science education. Some possibilities are listed under References for Additional Demonstrations and Activities (p. 20). In addition, magazines such as *Science and Children*, *Science Scope*, *Kid Science*, and *WonderScience*, often include ideas for demonstrations and classroom projects related to energy.

At this point, you should have already collected a variety of measures of student learning. This might include your observations of the students working in small groups, evaluations of group and individual data and observation sheets, journal or other directed writing assignments, and performance-based assessment for science process skills. Table 1 contains a rubric you might find useful when observing students during individual activities. If you use this rubric several times during the unit, you can monitor student progress toward meeting the goals listed. In addition, some of the individual activities have assessment sections that provide detailed suggestions for applying these general strategies to those specific activities. As a final unit assessment, an additional toy could be analyzed using the questions from the "Apply Your Energy Knowledge" activity. If you choose to give a more standard written examination as well, remember to ask some "why" or "explain" questions along with definitions and other objective-type questions.

Table 1: Assessment Table	
Criteria for evaluating student performance in an activity	Points
a. Demonstrates independence in the interpretation and performance of instructions.	
b. Demonstrates the ability to control variables.	
c. Demonstrates the ability to make accurate measurements.	
d. Demonstrates an understanding of the energy concepts involved in the activity.	
e. Analyzes the results and forms appropriate conclusions.	
f. Works effectively with team members.	

Performance Rating Scale:

No Evidence:	0 points	
Approaches Goal:	1 point	
Meets Goal:	2 points	
Exceeds Goal:	3 points	

REFERENCES

Additional Demonstrations and Activities

Bosak, S.V. *Science Is ...;* Scholastic Canada: Richmond Hill, Ontario, 1991.

Edge, R.D. *String and Sticky Tape Experiments;* American Association of Physics Teachers: College Park, MD, 1981.

Gartrell, J.E.; Schafer, L.E. *Evidence of Energy;* National Science Teachers Association: Washington, D.C., 1990.

Johnsey, R. *Problem Solving in School Science;* Macdonald Educational: London, 1986.

Liem, T.L. *Invitations to Science Inquiry,* 2nd ed.; Ginn: Lexington, MA, 1987.

Lowery, L.F. *The Everyday Science Sourcebook;* Dale Seymour: Palo Alto, CA, 1985.

Physics Fun and Demonstrations with Professor Julius Sumner Miller; Blasi, R.C., Ed.; Central Scientific: Chicago, IL, 1974.

Thier, H.D.; Knott, R.C. "Energy Sources"; *SCIS 3 Teacher's Guide;* Delta Education: Hudson, NH, 1992.

VanCleave, J.P. *Teaching the Fun of Physics;* Prentice-Hall: Englewood Cliffs, NJ, 1985.

Children's Trade Books

(Additional books are listed in the Further Reading section of each activity.)

Allison, L.; Katz, D. *Gee, Wiz! How to Mix Art and Science or The Art of Thinking Scientifically;* Yolla Bolly: Covelo, CA, 1983.

Craig, A.; Rosney, C. *The Usborne Science Encyclopedia;* EDC: Tulsa, OK, 1988.

Kent, A.; Ward, A. *The Usborne Introduction to Physics;* EDC: Tulsa, OK, 1983.

McCormack, A.J. *Inventors Workshop;* David S. Lake: Belmont, CA, 1981.

Ross, M.E. *What Makes Everything Go?;* Yosemite Association: Yosemite National Park, CA, 1979.

Spurgeon, R.; Flood, M. *Energy & Power;* Usborne: Tulsa, OK, 1990.

Walpole, B. *175 Science Experiments to Amuse and Amaze Your Friends;* Random House: New York, NY, 1988.

World Book's Young Scientist Series, Vol. 8; World Book: Chicago, IL, 1995.

Content Review

Energy is one of the most abstract concepts in physics, and it is also one of the most useful. Energy as an idea crosses all the artificial boundary lines we erect between studies of motion, electricity, heat, and optics, as well as those between the various branches of science, chemistry, physics, biology, geology, and others. Energy is a unifying concept. It forces us to look at nature in a holistic way. However, it is very hard to define other than mathematically.

CONSERVATION OF ENERGY

A common textbook definition of energy is the ability to do work, but this definition is circular because work is a way of transferring energy. Energy is a property of an object much like mass or volume, but it is more difficult to visualize. Because it is difficult to define, we will begin our discussion by talking about the characteristics of this important property. Energy is one of a small number of invariants—properties of nature that do not change under certain conditions—that help scientists to understand the physical world. Other invariants include electric charge, mass, and momentum. For energy, the conditions under which it does not change are very broad. The total energy of an isolated system never changes. (An isolated system is one in which no matter or energy can get in or out. See Figure 1.) This statement is often called the principle of conservation of energy, and we will begin our review of energy with this principle. (Note that the verb "conserve" here means "to remain constant," not "to save or preserve," as the term is more commonly used.)

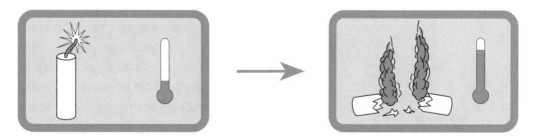

Figure 1: The total energy in an isolated system will always remain the same—it simply changes from one form to another.

Energy is a property of an object or a system of objects to which we can assign a numerical value. No matter what happens to that system, as long as it is not allowed to exchange energy with another system, this numerical value remains the same. Imagine a firecracker or a bomb inside a perfectly sealed, perfectly insulated box. We calculate the energy of the system (everything inside the box), and then the firecracker explodes. Now, when we calculate the energy of the system again, we will find that although the system now has changed many of its properties, its total energy is still the same number.

In order to aid this process of assigning a numerical value to energy, we categorize energy in different forms, which can change from one to another with the total always remaining the same. Nobel Laureate and master teacher Richard Feynman used to tell his students a story about Dennis the Menace in order to illustrate this idea. (See Feynman, pp 4-1– 4-2, or Kirkpatrick, pp 117–118.) In this story, Dennis the Menace spends each day playing in his room with 28 indestructible blocks. At the end of each day, his mother counts the blocks and discovers that no matter what Dennis uses the blocks for during the day, 28 always remain at the end of the day. Then one day she finds only 26 but notices that Dennis's toybox is heavier than normal. By carefully weighing the box and knowing the weight of one block, she is able to determine that two blocks are in the toybox. Another day she finds that blocks are missing but that the depth of the water in the bathtub is greater than it was in the morning. She cleverly determines a way to calculate how many blocks are invisible beneath the water. Over many days she learns that, if she is very careful to take into account all the blocks she can see, blocks in the toybox, and blocks in the bath water, there will always be 28 blocks. Then one day she counts the blocks and finds 30. After some investigation, she finds that one of Dennis's friends has been to visit and left two of his blocks. On another day she comes up short and finds that Dennis has thrown some of the blocks out the window. Now she must modify her rule of block counting. If she is careful to count all the possible places blocks can be and she is careful that no blocks enter or leave the room, then she will always have the same number of blocks.

To put this parable in energy terms, let each block represent a unit of energy and the different locations represent different forms that the energy can be in. Thus, we conclude that, if we add up all possible forms of energy and do not let any energy enter or leave our system, then we will always find the same total amount of energy. (See Figure 2.)

Figure 2: You can think of energy as a set of indestructible blocks. Although you may misplace some from time to time, they can never just disappear.

You may have noticed that energy still has not been defined, and it will not be given a general definition here, although we will define some specific forms of energy. You will find that as you and your students experience energy and energy transformations in many different contexts you will develop an intuitive sense of when an object has energy and when it does not. This forms a basis for more precise, mathematical definitions of various forms of energy in later grades and is quite sufficient for now.

ENERGY OF MOTION

The least abstract form of energy is kinetic energy, or energy of motion. Every moving object has kinetic energy. A famous physicist once described energy as "the Go of things" meaning that an object must have energy in order to move. The faster the object is moving, the greater its kinetic energy. Kinetic energy is proportional to the speed of the object squared, so a car moving at 60 mph has four times as much energy as it would have at 30 mph. (See Figure 3.) The amount of kinetic energy a moving object has also depends on the mass of the object. Kinetic energy is directly proportional to mass. A 2-kg object moving with the same speed as a 1-kg object will have twice as much kinetic energy. (See Figure 4.) Kinetic energy is not related to the direction the object is moving.

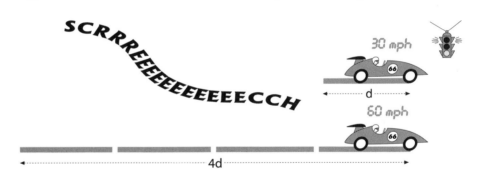

Figure 3: Stopping distance can be thought of as an indicator of the amount of kinetic energy (KE) that a vehicle possesses. Since $KE = \frac{1}{2}mv^2$, a car traveling at 60 miles per hour has four times the KE, and therefore requires four times the stopping distance (d), of a car traveling at 30 miles per hour.

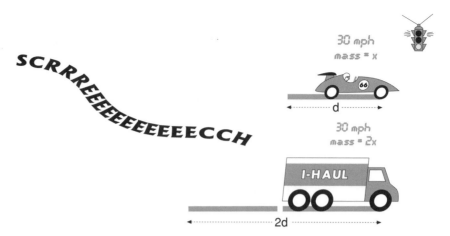

Figure 4: Doubling the mass will only double the total KE. A truck of mass 2x will have twice the KE, and therefore require twice the stopping distance (d), of a car with a mass of x (assuming both vehicles have comparable brakes and tires).

In the metric system, energy is measured in units of joules. This unit is named for James Prescott Joule, a British scientist who, in the nineteenth century, did important experiments to better understand the nature of energy. To calculate the kinetic energy of an object in joules, one uses the formula

$$KE = \frac{1}{2}mv^2$$

with mass in kilograms and speed in meters/second. Thus, a 2-kg object moving at 1 m/s would have 1 joule of KE. Perhaps an easier way to visualize what 1 joule of energy means is to think of it as the amount of energy it takes to lift an object that weighs 1 Newton (imagine a stick of margarine) a distance of 1 m. (We will soon see why this is so.) Thus, one joule is a fairly small amount of energy.

It is not suggested that your students actually do energy calculations, because it is difficult to account for all forms of energy involved in an experiment and therefore difficult to illustrate energy conservation quantitatively. Also, the joule is not an energy unit that students are likely to encounter in daily life, so the calculations would not help them connect the science lesson to the real world. Later in this review, some other possible energy units that they might have encountered will be discussed.

ENERGY THAT IS STORED

Anything that causes an object to move must be giving the object kinetic energy. This energy must come from somewhere. Consider a spring-loaded dart gun or a wind-up car. When released, these springs jump back to their normal shape, causing the dart or car to move. Energy is stored in the spring and later shows up as kinetic energy of the dart or car. This stored energy is called elastic potential energy (EPE). It is called "elastic" because objects that have this kind of energy are often stretchy, like springs and rubber bands, and "potential" because the spring has the potential to make the dart move even though nothing is happening right now. The numerical value associated with the potential energy in the spring depends on how much the spring is compressed and how stiff it is. The more the spring is compressed, the greater the amount of energy stored. Also, the harder the spring must be pushed to achieve a given compression, the greater the amount of energy stored.

Another way to get an object to move is to drop it. As it falls, it speeds up, so its kinetic energy is continually increasing. Where does this kinetic energy come from? An object positioned above the surface of the Earth so that it has the "potential" to fall is said to have gravitational potential energy (GPE). As the object falls, it loses gravitational potential energy and gains kinetic energy. The amount of GPE an object has is directly proportional to its weight (w) and its height (h) above the ground (GPE=wh). At the same height, an object with twice as much weight has twice as much energy. Likewise, an object has twice as much GPE when held 2 m above the ground as it has at 1 m above the ground. If we hold an object weighing 1 Newton (remember your stick of margarine) 1 m above the ground, it has a gravitational potential energy of 1 joule. In the previous section, we said it took 1 joule of energy to lift it up 1 m. Now this makes sense, since you must supply the energy that gets stored as gravitational potential energy.

While we usually speak of gravitational potential energy as belonging to the object being dropped, it is actually a property of the system composed of the object and the Earth. If the Earth weren't there to pull the object down, the "potential" to move would not exist. (Thus, it makes sense that gravitational potential energy depends on the object's weight rather than its mass.) We usually associate this form of energy with just the object because when the object is dropped, it is the object that gains kinetic energy, not the Earth.

Having been taught that, neglecting air resistance, all objects fall at the same rate independent of their weight, some students find the idea that gravitational potential energy depends on the weight of the object confusing. The following example might be helpful. If a soccer ball and a bowling ball are both dropped from a distance of 1 m above the floor, they will hit the floor traveling at the same speed. However, the bowling ball is much more likely to damage the floor. Because the weight of the bowling ball is much greater than that of the soccer ball, the bowling ball has more energy that it can transfer to the floor. It had more GPE before it was dropped, so it has more kinetic energy when it strikes the floor. Imagine

Figure 5: A soccer ball and bowling ball fall at the same rate of speed.

that you are the floor: wouldn't it hurt more to have a bowling ball dropped on you than a soccer ball? What if you placed a tomato under the bowling ball and under the soccer ball? (See Figure 5.) Which ball will have the greater effect on the tomato under it? Clearly speed is not the only important variable in this case.

Many simple examples can be used to illustrate changes between these three kinds of energy (which together are called mechanical energy). Think of a roller coaster car at the top of the first hill. It has a great deal of gravitational potential energy, because it is very high above the ground. As it goes down the hill, its GPE is gradually changed into kinetic energy. At the bottom of the hill, almost all its initial GPE has been changed into kinetic energy, and the car is traveling quite fast. As it starts up the next hill, it begins to slow down, and its kinetic energy is gradually turned back into gravitational potential energy. In order for the car to make it over the second hill, the hill must be lower than the first one, so that when it reaches the top it still has some kinetic energy left. If the second hill were taller than the first, then, when the car reached the height of the first hill, all its energy would be converted back to GPE and it would have no kinetic energy. It could not continue to move forward; in fact, it would instead roll back down the hill, gaining kinetic energy again. (See Figure 6.)

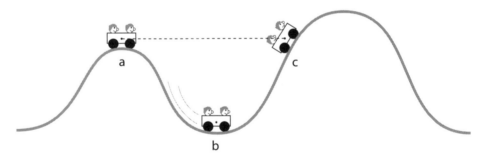

Figure 6: A badly designed roller coaster. At point a, the car has a certain amount of gravitational potential energy (GPE). As the car rolls down the hill the GPE is changed into kinetic energy (KE). At point b, the car has maximum KE and minimum GPE. At point c the car has the same GPE that it began with (discounting any losses due to friction or air resistance) and cannot go any farther unless energy is added from outside.

A simple pendulum is also a good example of energy conversion. (For convenience, let us assume for the moment that the pendulum is frictionless.) To get it started, pull it to one side (which lifts it up) giving it gravitational potential energy. When you release it, the pendulum swings down, converting the GPE to kinetic energy. As it swings back up on the other side, the kinetic energy returns to GPE. It will swing back up to exactly the same height it started from, because it must have the same amount of GPE at the end of the swing as the beginning. As it continues to swing, energy transforms back and forth between the two forms. (See Figure 7.) As you know, the pendulum will eventually stop due to friction, but we'll come back to that a little later.

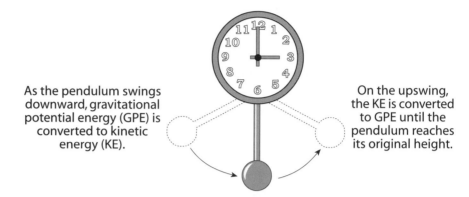

As the pendulum swings downward, gravitational potential energy (GPE) is converted to kinetic energy (KE).

On the upswing, the KE is converted to GPE until the pendulum reaches its original height.

Figure 7: Gravitational potential energy is converted to kinetic energy in a frictionless pendulum.

A similar example using elastic potential energy instead is a block attached to a horizontal spring oscillating back and forth on a smooth (frictionless) table. Stretch the spring to start the motion. This stretching stores energy in the spring. When the spring is released, it returns to its normal length, causing the block to move. The elastic potential energy stored in the spring is converted into kinetic energy of the block. The block is moving when the spring returns to its normal length, so the block continues to move, compressing the spring. The spring exerts a force on the block, slowing it down and returning the kinetic energy to EPE in the compressed spring. (See Figure 8.) As with the pendulum, the system will continue to oscillate back and forth. If we hang the spring vertically, then both gravitational potential energy and EPE are involved in the system, since the block would be changing its height above the ground while the spring stretched.

The stretched spring has elastic potential energy (EPE).

a block on a smooth surface

As the spring contracts, EPE is converted to kinetic energy (KE).

KE is converted back to EPE as the spring compresses.

Figure 8: A block attached to a spring oscillates back and forth on a smooth table.

WORK: A WAY TO TRANSFER ENERGY

Let's return to the subject of how an object gets kinetic energy. One way to give an object kinetic energy is to push it. This push is a direct conversion from energy in your body (discussed later) to kinetic energy; it does not involve storing the energy up in an intermediate step. One way to think about this is to say that we have done work on the object. Work is one mechanism for transferring energy from one object to another. Work always involves the use of a force. The simplest case in which to calculate the work done by a force is the case in which the size and direction of the force are constant and the object moves in the direction of the applied force. In this case, the numerical value of the work done is the force multiplied by the distance that the object moves. It may seem at first that the object should always move in the direction of the applied force, but this may not be true if the object is already moving in a different direction or if more than one force is being exerted on the object. For instance, think about a child pulling a little red wagon. The child is exerting a diagonally upward force on the handle and the wagon is moving horizontally. In this case, several other forces act on the wagon in addition to the child's pulling force: gravity, friction and the upward force of the ground on the wheels. (See Figure 9.)

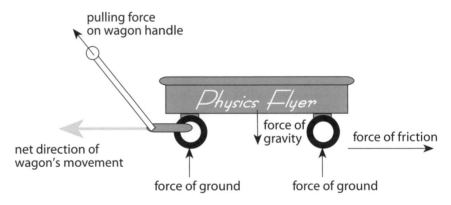

Figure 9: All of the forces add up to result in horizontal motion.

Now let's return to some of our previous examples. The spring in the dart gun stores energy that it later gives to the dart. Where did that energy come from? Someone exerted a force on the spring and caused it to compress. That person did work on it. Likewise, when we lift a ball before letting it fall, we do work on it to give it gravitational potential energy. The roller coaster car had to be dragged up the hill by some mechanical system that exerted a force on the car, doing work in the process.

Not only can work be done to give an object kinetic energy, but an object that has kinetic energy can do work. A moving hammer can hit a nail, exerting a force on the nail and causing it to move. A moving block can compress a spring, storing up energy for later use. A moving car can hit a stationary object, exert large forces, and cause great damage.

A practical example that brings together everything we have discussed so far is the operation of a hydroelectric plant. In a hydroelectric plant, water moves from the top of the dam to the bottom of the dam, losing gravitational potential energy in the process. This GPE is changed into kinetic energy. When the water hits the turbines, the force does work, changing some of the water's kinetic energy into kinetic energy of the moving turbines. (See Figure 10.) The kinetic energy of turbines eventually is converted to electrical energy. (Exactly how this conversion happens is beyond the scope of this module.) Some hydroelectric plants utilize a process called pumped storage. During the night, while the demand for electricity is low, some of the electrical energy produced is used to pump water from below the dam back up into the lake. This stores energy in the water in the form of gravitational potential energy. Then, during peak demand hours, the water is again allowed to flow to the bottom of the dam, repeating the cycle of GPE to kinetic energy to electrical energy. Pumped storage is about 80 percent efficient, meaning that about 80 percent of the energy used to pump the water back up is actually stored in the water, so about 20 percent of the energy is lost. However this percentage still compares favorably with other losses that occur through wide fluctuations in power demand. In other words, the plant itself is much more efficient if it can maintain a constant output.

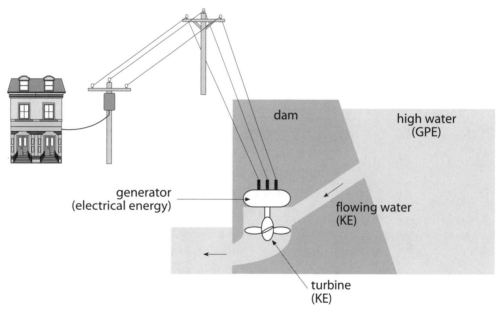

Figure 10:The conversion of gravitational potential energy to kinetic energy to electrical energy in a hydroelectric dam.

OTHER FORMS OF ENERGY

Energy transformations go on constantly. The previous section contains examples of the transformation of kinetic energy to potential energy and vice versa. However, sometimes KE seems to disappear. A car coasts to a stop; its original KE is gone. Our pendulum eventually stops oscillating, as does our block-spring system. In cases like this, thermal energy is always produced—something gets warmer. The increase in temperature is a macroscopic manifestation of the fact that molecules are moving faster.

Consider a bowl of water. As a whole it is not moving; however, its individual molecules are moving randomly in all directions. Its thermal energy is the sum of the KE of all of its molecules. If energy is added to the water, its molecules will move faster and an increase in temperature can be measured. Whenever KE appears to be lost, it has almost always been converted into thermal energy. Note that the term "heat" has not yet been used. Heat is a transfer of energy just like work. Heat is a transfer of energy between two objects that takes place because of a temperature difference between the objects. If an ice cube is placed in a bowl of water, heat will move from the water to the ice, because the water is at a higher temperature. On the other hand, the soles of your shoes get warm when you run not because the ground is hotter than your shoes and transfers heat to them, but because the frictional force between the shoes and the ground increases the thermal energy of the molecules in the shoes. While the distinction between heat and thermal energy is important to physicists, it is probably not important that your students understand it, so you should feel comfortable using heat for both meanings.

Almost all processes found in nature involve transformations of energy from one form to another. Of course, energy exists in many more forms than the three we have discussed so far. Electrical energy, which we mentioned in the hydroelectric dam example, is related to the electric forces that charged particles exert on one another. Let's think about a simple circuit in which a battery causes a light bulb to glow. The battery and bulb are connected to one another with metal wires, and the light bulb has a metal filament. An important characteristic of metals is that some of the electrons of the atoms are not tightly bound to individual atoms and can move around easily. The chemical reaction inside the battery causes positively charged particles to accumulate at one electrode of the battery and negative ones at the other electrode. These internal electrodes are connected to the external positive and negative terminals. These charges exert forces on the electrons in the wires. These forces cause the electrons to move and thus to have kinetic energy. As they move through the wire, they frequently interact with the stationary atoms in the wire, giving them some of their energy.

Although the real interaction is more complicated, you can liken this process to the action of billiard balls bumping into each other. The atoms are not free to travel as the electrons do, so the extra energy from the moving electrons causes them to vibrate more. Thus, we have changed the kinetic energy of the moving electrons into thermal energy of the atoms.

This increase in thermal energy can be detected as an increase in the temperature of the wire. The filament is made of a kind of wire that gets very hot when electrons pass through it. (Its properties allow it to get very hot without vaporizing.) The filament gets so hot that it glows just like a hot stove burner. The light that is produced is another form of energy. In summary, the battery exerts an electrical force on the electrons in the wire, giving them kinetic energy. As these electrons bump into the atoms, their kinetic energy is changed to thermal energy. Sometimes the atoms accumulate so much energy that they get rid of some of it by giving it off in the form of light. This is the basic principle on which light bulbs work.

Another common form of energy is chemical energy. The chemical energy stored in gasoline runs your car just as the chemical energy stored in the food you eat provides energy for your body. Chemical energy is a type of potential energy associated with the positions of the charged particles in atoms and molecules. (Remember that gravitational potential energy is related to the position of an object relative to the Earth.) When atoms or molecules are rearranged during chemical reactions, some of that stored energy may be released. For instance, when coal is burned, carbon atoms in the coal combine with oxygen molecules in the air to produce carbon dioxide. In this process energy is released. One way to think about this is that in order to take the carbon dioxide apart again into carbon and oxygen, we would have to do work or put energy into the system. Thus, the system must have lost energy when the original atoms combined. All chemical fuels, including food, provide energy in a similar manner.

The energy we receive from the sun is in the form of radiant energy (sometimes called light energy). This is the energy carried by electromagnetic waves. The light that we see is just one kind of electromagnetic wave. Radio waves, x-rays, infrared light, ultraviolet light, and microwave radiation are also electromagnetic waves. The sun emits some of all of these kinds of waves, but visible light, infrared light and ultraviolet light are the most important in terms of bringing energy to the Earth. All waves can carry energy. A water wave created by a boat in one part of a lake can make a Styrofoam™ cup floating in another part of the lake bob up and down. When we spoke above about the atoms in a light bulb filament emitting light, we also could have said that they produce electromagnetic waves.

The sun's energy and the electrical energy produced by nuclear power plants start out as nuclear energy. Nuclear energy is related to the force that holds atomic nuclei together. Consider a large atom such as uranium. When the atom was formed, large forces were required to force the protons together into the nucleus. This is due to the fact that the protons are all positively charged, and thus they repel one another. (You may be asking how these large nuclei manage to stay together in the first place. Once the protons and neutrons are very close together, they are held by a force called the strong nuclear force. It is a much larger force than the electrical force pushing the protons apart; however, it has a very short range, so the protons must be very close together before it can hold them.) Some of the energy that went into bringing the protons together is stored as potential energy in the nucleus. When the uranium atom later breaks apart into two smaller atoms as it does in the fission process, some of this potential energy is turned into kinetic energy of the new atoms. A nuclear reactor is designed so that this kinetic energy can be used to change liquid water into steam. The steam is then used to turn a turbine to generate electricity, just as in a coal-fired power plant.

The energy generated in the sun is created in a different nuclear process called fusion. It turns out that for small atoms, energy is released when two atoms are combined into one. Given the discussion in the previous paragraph, this doesn't seem possible, but it is. We still have to put energy into the system to get the atoms close enough together that the strong nuclear force can hold them together. (In order for a nuclear power plant to produce energy using fusion, the gas of hydrogen atoms would have to be heated to a very high temperature.) But once we get several hydrogen atoms to combine to make one helium atom, the strong nuclear force holds the particles in the nucleus so tightly that excess energy is released. At different stages in a star's life, hydrogen, helium, carbon, oxygen, and other light mass elements are all involved in the fusion process.

We sometimes also speak of sound energy, because it is convenient to link the energy to the macroscopic result. (Macroscopic means large-scale—the opposite of microscopic.) However, sound energy is really just the kinetic energy of the moving air molecules that carry the sound wave.

Your students may be familiar with energy units that are often used for some of these other forms of energy. BTU (an abbreviation for British Thermal Unit) is a unit of energy often used when measuring heat. Heat-producing appliances are often rated in BTUs per hour. The calorie is another energy unit often used for heat. The term used to describe food energy is "Calorie" (with an upper-case C), which is equal to 1,000 calories (with a lower-case c). When we speak of the number of Calories in food, we are talking about the amount of energy your body will get when it breaks down the molecules. Electrical energy is often measured in kilowatt-hours (kwh). It would perhaps be more sensible if all forms of energy were measured in the same units, but for a variety of reasons they are not. For instance, the calorie was defined as a unit of heat before anyone realized that heat was a form of energy.

Let's return to the principle of conservation of energy: In an isolated system, energy may change from one form to another, but the total energy in the system is always constant. The total energy in the universe is a constant. Occasionally in the history of science, circumstances have developed in which it seemed that energy was not conserved. In all cases, it has turned out that energy was not lost but rather turned into a new form we had not known about before.

THE ENERGY CRISIS

Conservation of energy seems to be in conflict with one of modern society's biggest worries—the energy crisis. The crisis is that we are using up our energy supply and sometime in the future we may run out of energy. But if the total amount of energy is constant, how can we run out? We are not "using up" energy, but we are changing it from forms that are easy for human beings to use into forms that are hard to use. It is easy to burn coal and use the released chemical energy to create electrical energy. It is easy to use that electrical energy to heat our homes and cook our food. However, once that energy is turned into heat, it spreads out throughout the atmosphere and is virtually impossible to gather back up and change into some other form of energy again. For all practical purposes, it is lost. It is easy to use the chemical energy in gasoline to give your car kinetic energy, but eventually all of the car's kinetic energy winds up as heat energy when you use the brakes to stop the car. While the same quantity of energy remains, you cannot take that heat and turn it back into kinetic energy of the car. This is how it is possible to have an energy crisis in a world in which energy is conserved.

REFERENCES FOR FURTHER CONTENT REVIEW

Faughn, J.S.; Turk, J.; Turk, A. *Physical Science;* Saunders: Philadelphia, PA, 1991.

Feynman, R.P.; Leighton, R.B.; Sands, M.L. *The Feynman Lectures on Physics;* Addison-Wesley: Reading, MA, 1963; Vol. 1.

Hazen, R.M.; Trefil, J. *Science Matters;* Doubleday: New York, NY, 1991.

Hewitt, P.G. *Conceptual Physics,* 7th ed.; HarperCollins: New York, NY, 1993.

Hewitt, P.G.; Suchocki, J.; Hewitt, L.A. *Conceptual Physical Science,* HarperCollins: New York, NY, 1994.

Kirkpatrick, L.D.; Wheeler, G.F. *Physics: A World View*, 2nd ed.; Saunders: Philadelphia, PA, 1995.

Priest, J. *Energy: Principles, Problems, Alternatives;* Addison-Wesley: Reading, MA, 1991.

✔ *Concept Introduction*

Concept Application

Synthesis

What Makes It Go?

...Students are introduced to the concepts of potential energy and energy conversion and apply these concepts to a familiar toy.

Push-n-Go toy

✔ **Time Required**

Setup	none
Performance	75 minutes
Cleanup	5 minutes

✔ **Key Science Topics**

- energy conversion
- force
- friction
- gears
- kinetic energy
- mass
- potential energy
- work

✔ **Student Background**

Students should know that an ordinary stationary object, such as a book, moves in the direction of the force when a force acts on it.

✔ **National Science Education Standards**

Science as Inquiry Standards:

- Abilities Necessary to Do Scientific Inquiry

  Students use observation and measurement to investigate the relationship between the mass of the toy, its initial speed, and its stopping distance.

  Students propose possible explanations for the relationships they have discovered.

Physical Science Standards:

- Transfer of Energy

  Energy is a property of an object which is associated with its motion, may be stored temporarily, and can be transferred from one object to another through a process known as work.

✔ **Additional Process Skill**

- hypothesizing

  Students hypothesize about why the toy does not move in the same direction as the force applied.

MATERIALS

For Getting Ready
Per class
- (optional) Phillips-head screwdriver

For Introducing the Activity
Per class
- ball

For the Procedure
Part A, per small group
- Push-n-Go® toy

The Push-n-Go by Tomy America is available at many toy stores, is inexpensive, and comes in a variety of shapes that have nearly identical internal mechanisms. The styles include a fire engine, dump truck, airplane, dinosaur, and others. It is marketed as a preschool toy, so many of your students may remember having one when they were younger. Some of them may be able to bring one from home for a few days. A number of other companies make toys with similar inner mechanisms. Press-N-Go™ Fire Engine by Shelcore®, Press-N-Go Cookie Monster by Illco® and Go-Go Gears™ by Playskool® are examples. The Push-n-Go comes apart somewhat more easily, but the others could also be used in this lesson.

- Phillips-head screwdriver
- small cup

Part B, per class
- (optional) book, ball, or toy car

Part B, per group
- variety of coins, rocks, or small weights, preferably fairly flat

If groups are measuring mass, the weights don't need to be identical. If they are not measuring mass, each group needs identical weights with a total mass equivalent to about 12 quarters.

- meterstick or measuring tape
- tape
- (optional) balance or other device to measure mass
- (optional) empty box and books

For the Extensions

❶ All materials listed for the Procedure plus the following:
Per class
- access to several types of horizontal surfaces, at least 1 foot x 10 feet, such as the following:
  - tile
  - linoleum
  - wood
  - ceiling tile
  - asphalt
  - sandpaper
  - grass
  - short-nap carpet
  - long-nap carpet
  - foam rubber pad

 One way to get carpet or foam rubber is to go to a carpet store and ask if they have any ends of rolls or other long, narrow scraps.

❷ Per class
- other toys in which energy is changed from one kind to another

SAFETY AND DISPOSAL

Depending on the age of your students, you may choose not to have them handle the screwdrivers and take apart the toys themselves. In this case, disassemble several toys beforehand, but keep one together to demonstrate. No special disposal procedures are required.

GETTING READY

You may want to loosen all the screws and then retighten them to make sure that none are stuck. This is more important with younger children.

INTRODUCING THE ACTIVITY

Gently roll a ball away from you. Have a student roll it back. Listen to students' observations and help them to conclude that you pushed away from yourself, and the ball moved away from you. Review the rule that when a force acts on a simple stationary object, the object moves in the direction of the force.

PROCEDURE

Part A: Exploring How the Push-n-Go Works

1. Demonstrate the Push-n-Go by pushing down on the rider's head and releasing it. Ask a student to demonstrate it a second time. Ask how this is different from the ball demonstration you just did. Listen to students' observations and help them to conclude that this time the force was downward and the movement was forward. Establish that, in this case, the object did not move in the direction of the force applied by the hand.

2. Ask the students to hypothesize about why the toy did not move in the direction of the force. Listen and restate their hypotheses, helping them to clarify their ideas. Guide the group to develop the hypothesis that something is being stored and then later causing the toy to move in a different direction. Tell students that the "something" that is being stored is called energy and that they will be learning about energy over the next several weeks. Ask the students to suggest what might be inside the toy that stored the energy and changed the direction. Have them record suggestions.

3. Tell the students that they are going to take the toys apart to check out their hypotheses. Divide the class into groups that are small enough for everyone to see the toy. Direct each group to use the screwdriver to remove the screws from the bottom of the Push-n-Go. Have each group put the screws in their small cup. Everyone should have an unobstructed view of the inside of the toy.

4. Have a student in each group push the rider down on the spring. Tell the groups to watch what happens, discuss what is occurring, and write a description of these events in their own words. Make sure they explain both how the energy is stored and how the direction of the force is changed. Encourage students who see and clearly understand what is happening to explain it to those who may be looking but not understanding.

5. Ask the students to discover how the gears are prevented from turning while the spring is being pushed down. Ask why this is necessary.

6. Through discussion, bring the class to consensus on how the toy stores energy, changes the direction of the force, and keeps the gears from turning while the head is being pushed down. (See the Explanation.)

7. Using the Push-n-Go as an example, introduce the terms "work," "kinetic energy," and "potential energy" as names for the concepts the class has been discussing. You may want to bring in additional examples of toys or other devices to which the concepts apply (see Content Review), but do not expect a full understanding of these terms at this time. Students will get lots more practice with these concepts as you move through the unit.

8. Have the students reassemble the toys, run them again, and practice narrating what is happening inside the toy as they use it. Explain to the students that their homework assignment is to share the toy with a parent or other adult and explain how it works.

9. Set up a schedule for each student who does not own a Push-n-Go to take the toy home. If desired, give each student a Take-Home Verification Form (provided) to take home for the signature of a friend or family member who has listened to the explanation and observed the toy.

 Your students will learn a lot from practicing, focusing on, and delivering the explanation. If toys have been brought in by students, be sure to check with parents before loaning these toys out to other students.

Part B: Investigating the Effect of Changing Mass on the Push-n-Go

1. You could introduce this experiment by pushing a book, ball, or toy car and asking, "How might we vary the initial speed at which an object travels?" Students are likely to answer that pushing it harder or more gently would vary the initial speed. "Pushing it harder—or more gently—does not work for the Push-n-Go. What would work?" Through class discussion, the idea should emerge that varying the mass of the Push-n-Go will alter its speed. Ask the students if they think the total distance traveled will also change.

2. Remind your students that in a well-designed experiment, we examine only one variable at a time. Ask students, "What is the variable in this experiment?" *The variable is the mass of the toy.* Students should remember to keep everything else constant, such as the specific Push-n-Go used and the surface on which it runs. Tell students that in this case, while they are changing one variable (mass), they are looking for its effect on two properties (speed and distance).

   Have each group of students write a hypothesis at the top of the Part B Data Sheet (provided). A simple hypothesis such as, "The more massive toy will move more slowly," or "The more massive toy won't go as far," is fine. If students wish to hypothesize about the actual distance change for a certain amount of extra mass (for example, 1 foot less if carrying four quarters), encourage them to do so.

3. Tell the students that as they carry out the experiments described in steps 4–6 they should carefully observe the initial speed of the Push-n-Go. They will not actually measure the speed but should be able to determine from observation whether the speed changes as the mass increases.

4. Have each group establish a baseline distance for the regular Push-n-Go by finding its mass, running it three times, measuring the distance, and averaging. Have them record the results on the Part B Data Sheet.

> *Determining the mass is not absolutely necessary. If you use identical objects such as quarters in step 5, then you could just count the number of quarters.*

5. Have each group load some weights, coins, or small rocks into the toy or tape them to the outside of the toy. Have each group find the mass of the entire object now (if desired) and record the results. Each group should run the toy three times, measure the distance each time, and record the results on the Part B Data Sheet.

6. Have each group repeat step 5 at least three times using different amounts of weights, coins, or rocks than before.

7. Have the students graph the results on the line graph on the Data Sheet, with the mass (or number of quarters) as the horizontal (x) axis and distance traveled as the vertical (y) axis. (See sample data in Figure 1.)

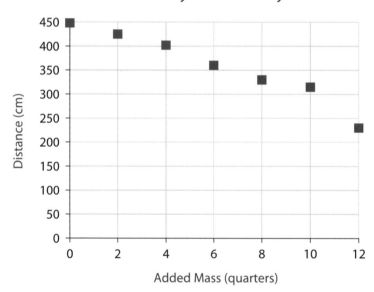

Figure 1: Effect of Adding Mass on Distance Traveled by Push-n-Go Toy

8. Ask the students what they observed about the speed of the toy during their experiments. Lead them to see that their observation makes sense in terms of the definition of kinetic energy. If two objects have the same amount of kinetic energy, the more massive one will be moving slower.

9. Ask students if it seems reasonable to them that the heavier the toy, the sooner it stops. Ask them why they think so. Someone may say that it is because there is more friction. You might want to illustrate this by letting someone push an empty box across the floor and then repeat the process with the box loaded with books. Tell the students they will return to the connection between friction and energy later.

EXPLANATION

➤ *The following explanation is intended for the teacher's information. Modify the explanation for students as required.*

When you operate the Push-n-Go, you apply a downward force to the rider's head, and it moves downward as a result. Thus, an applied force has moved through a distance, and work has been done. As you push, the head pushes back on your finger, showing that a force is being exerted. Clearly, the movement of the head downward occurs while the force is being applied.

Underneath the rider is a rather stiff spring. As the rider moves down, the spring compresses, storing energy. Once the head goes as far down as possible, nothing happens until you remove your finger. When you are explaining the toy in step 7, push down on the rider's head and hold your finger still while explaining that your work was converted into potential energy that is now stored in the spring. Your body's energy was transferred, through the process of doing work, to the spring.

When the rider is released, the spring extends, and it pushes the rider back up. As the rider rises, a rack (series of teeth or notches) on the rider exerts a force to turn a gear (called a pinion) that is mounted on an axle. (See Figure 2.) The pinion transfers this force through all the other gears to the wheels, causing the Push-n-Go to roll forward. You can turn the axle by hand to see the gears move in slow motion. The removal of one additional screw (or two, depending on the model) on the side of the gear housing detaches it from the base of the truck, making the lower two gears more easily visible.

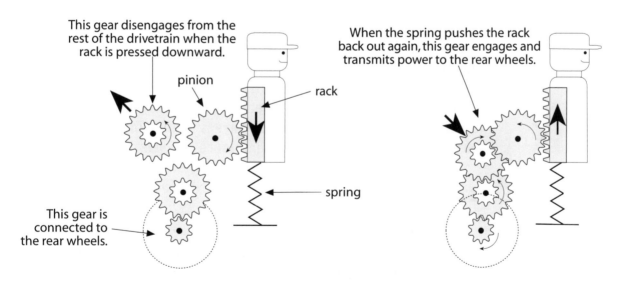

Figure 2: When the rider is pushed down, the spring compresses, storing energy. When the rider rises, a rack on the rider causes the gears to turn. (This figure shows the mechanism of the locomotive. The mechanisms of the other forms of this toy may be arranged slightly differently.)

The process can be described in energy terms as follows: the potential energy of the spring is converted into kinetic (or motion) energy of the gears and the toy as a whole. A small part of the energy is converted to heat energy by friction between the parts.

Since the rack causes the axle to turn while the rider is being pushed up, why doesn't the same thing happen when the rider is being pushed down? The motion of the axle while the spring is being compressed is prevented by an interesting feature of one of the middle gears. This gear is mounted in a slot that allows about half a centimeter of vertical travel. When the spring is extending, the pinion pushes down on this middle gear, causing it to turn and to make contact with the gear below it; but when the spring is being compressed, the pinion turns in the opposite direction and lifts up on the middle gear, causing it to move away from the gear below it. Thus, no contact is made between the middle gear and the gear below, and the axle does not turn.

When the Push-n-Go stops rolling, you might think the energy has disappeared. However, the energy is not lost. Rather, it has changed into thermal energy because of the friction between the wheels and the floor. On a carpeted floor, where there is a greater amount of friction, the Push-n-Go will not travel as far as on a hard floor before all the kinetic energy has been changed to thermal energy. If two students send the Push-n-Go rapidly back and forth, the rubber rings on the wheels will begin to feel warm. This observation helps students verify that the kinetic energy has not been lost but changed into thermal energy.

In Part B of the Procedure, students observe the effects of increasing the mass of the toy: lower initial speed and shorter distance traveled. The potential energy stored is the same in all trials. Thus, the initial kinetic energy of the toy is also constant. Since kinetic energy is proportional to both mass and speed, as mass increases speed must decrease. The distance the toy can travel before all the kinetic energy is converted into thermal energy depends on how much kinetic energy the toy has initially (which is constant) and how large the friction force is. Adding mass increases the friction force and thus shortens the distance traveled.

EXTENSIONS

1. Surfaces Extension

a. Remind students that friction between the toy and the floor caused the toy to stop running after it had gone about 5 m. Kinetic energy was changed into heat energy. Ask them, "What would happen on a higher-friction surface? What would happen on a lower-friction surface?"

b. If your students have had sufficient experience, ask them to design and carry out an experiment to determine which of several surfaces exerts the largest friction force on the toy. If they have not had sufficient experience designing experiments, use steps c–e to lead them through the design and testing process step by step.

c. Explain to the students that they will be comparing the way the Push-n-Go runs on several different surfaces with differing amounts of friction by measuring the distances traveled by the toy on each surface.

d. Ask the students to choose several surfaces from those that you have available to use in the experiment.

e. Remind the students that in a good experiment only one variable changes; all other factors that could affect the results are held constant. Ask students, "What is the variable in this experiment?" *The variable is the surface.* Explain that since some Push-n-Gos may run better or worse than others, the same Push-n-Go should be used for all the work by one group.

Tell students that another feature of a good experiment is a hypothesis. Have each group write a hypothesis on the line at the top of the Extension Data Sheet (provided). This hypothesis should state which surface they think will produce the least amount of friction and therefore will allow the Push-n-Go to travel the farthest.

f. Have each group run their toys several times on each of the surfaces they have chosen to test. Have students record the distances traveled on the chart, compute averages, and plot a bar graph. (See sample data in Figure 3.)

Figure 3: Average Distance Traveled by Push-n-Go Toy on Different Surfaces

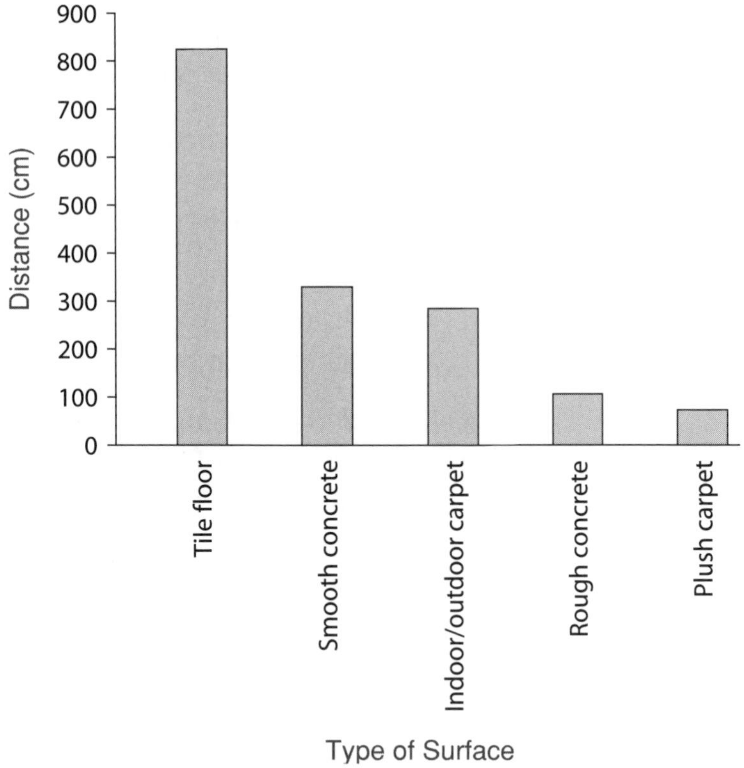

g. Have each group list the surfaces they tested in order of longest roll to shortest roll. Ask the students, "Do all results agree? If not, what might have caused the difference?" Discuss the results in terms of the size of the friction force exerted by each surface.

2. Energy Conversions Extension

Ask students to give examples of other toys in which energy is changed from one kind to another. You might ask the students to bring in toys the next day that illustrate energy conversion. Many toys fit this criterion. All toys using springs, whether compressed or wound, could be analyzed in a manner similar to the Push-n-Go, as could even simpler toys such as paddleballs and bouncing balls. Any toy that uses batteries converts electrical energy into kinetic energy, light energy, sound energy, or thermal energy.

ASSESSMENT

Options:

• Informally observe students during performance of the activity and during class discussion. You may also collect and evaluate the Part B Data Sheet.

- For an activity-based evaluation, students could examine the workings of the manual pencil sharpener in the classroom and describe how muscle energy is converted to sharpen a pencil. Be sure to instruct students to include in their explanations the workings of the pencil sharpener's gears and grinders.

CROSS-CURRICULAR INTEGRATION
Language arts:
- Have students imagine the following scenario: Push-n-Go toys have been popular for a number of years because of their unique design and durability, but recently sales have slumped. You have the following assignment as an advertising agent: Describe your designs for the new and improved Push-n-Go toy line. Include in your discussion: 1) why you think this design will sell; 2) who your targeted buyers are; and 3) a possible price for the product. Finally, make up an advertisement for the new design and include illustrations.

Math:
- Use this activity as a math activity, focusing on measuring, graphing, and charting, with science as the secondary focus.

FURTHER READING

Faughn, J.; Turk, J.; Turk, A. *Physical Science;* Saunders: Philadelphia, PA, 1991. (Teachers)

Gartrell, J.E.; Schafer, L.E. *Evidence of Energy;* National Science Teachers Association: Washington, D.C., 1990. (Teachers)

Kirkpatrick, L.D.; Wheeler, G.F. *Physics: A World View,* 2nd ed.; Saunders: Philadelphia, PA, 1995. (Teachers)

HANDOUT MASTERS

Masters for the following handouts are provided:
- Take-Home Verification Forms
- Part B Data Sheet
- Extension Data Sheet

Copy as needed for classroom use.

What Makes It Go?

Take-Home Verification Forms

- -

_____ showed me the Push-n-Go toy and explained
to me how it works on the inside.

Name of Student

Signature   _____

Date   _____

- -

_____ showed me the Push-n-Go toy and explained
to me how it works on the inside.

Name of Student

Signature   _____

Date   _____

- -

_____ showed me the Push-n-Go toy and explained
to me how it works on the inside.

Name of Student

Signature   _____

Date   _____

- -

_____ showed me the Push-n-Go toy and explained
to me how it works on the inside.

Name of Student

Signature   _____

Date   _____

- -

 Reproducible page from *Exploring Energy with **TOYS*** published by Terrific Science Press™

What Makes It Go?
Part B Data Sheet

Form a hypothesis about the effect of changing the mass of the toy:

_____

_____

_____

Chart

Trial Number	Distance Traveled				
	No extra load _____ (Mass)	Extra load 1 _____ (Mass)	Extra load 2 _____ (Mass)	Extra load 3 _____ (Mass)	Extra load 4 _____ (Mass)
1					
2					
3					
Average					

Graph

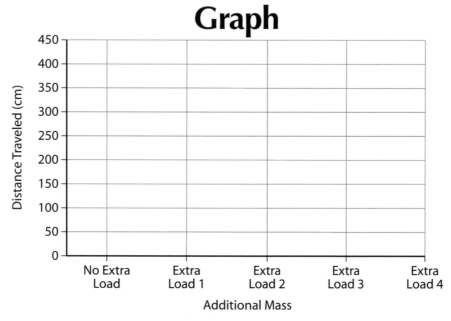

Name _____ Date _____

What Makes It Go?
Extension Data Sheet

Form a hypothesis about which surface will produce the least amount of friction and how that will affect the distance the toy travels:

_____

_____

_____

Chart

Trial Number	Distance Traveled			
	Surface #1: _____ (Write name of surface.)	Surface #2 _____ (Write name of surface.)	Surface #3 _____ (Write name of surface.)	Surface #4 _____ (Write name of surface.)
1				
2				
3				
Average				

Bar Graph

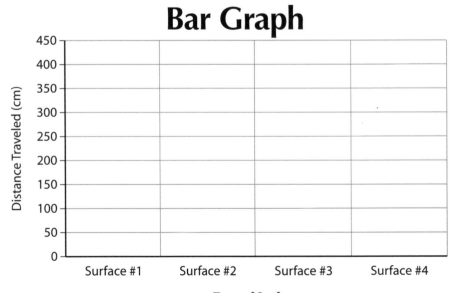

☑ *Concept Introduction*

Concept Application

Synthesis

The Toy That Returns

...Explore the concepts of potential energy and kinetic energy using rubber bands as "storers" of energy.

✔ **Time Required**

Setup   15 minutes
Performance 40 minutes
Cleanup   5 minutes

✔ **Key Science Topics**

- kinetic energy
- motion
- potential energy
- work

✔ **Student Background**

Students should have already been introduced to the concept of kinetic energy and potential (or stored) energy. This activity is a good follow-up to the activity "What Makes It Go?"

Commercial and homemade come-back toys

✔ **National Science Education Standards**

Science as Inquiry Standards:

- Abilities Necessary to Do Scientific Inquiry

  Students build come-back toys and observe their motion.

  Students write explanations of how the toys work based on their observations.

Physical Science Standards:

- Transfer of Energy

  Energy is a property of an object that is sometimes associated with its mechanical motion.

  Energy can be transferred repeatedly from kinetic (motion) energy to potential (stored) energy in a rubber band and then back to kinetic energy.

MATERIALS

For Getting Ready
Per class
- can opener
- electrical or masking tape
- scissors or knife with pointed tip

Per student (ideally) or per group of no more than 4 students (plus 1 set for teacher)
- small (1 pound) coffee can

 Larger cans can be used but will require larger rubber bands. Have students start collecting cans and plastic lids several weeks in advance.

- 2 plastic lids for the coffee can

For the Procedure
Per class
- commercial come-back toy such as Come Back Roly Toy by Shackman, Mickey Mouse Roll Back Wheel by Illco™, or Rollback Pals™ by Child Dimension™

Per student (ideally) or per group of no more than 4 students (plus 1 set for teacher)
- coffee can prepared in Getting Ready
- 2 plastic lids prepared in Getting Ready
- rubber band about 8–10 cm long (unstretched)

 Be sure to have some extras available to replace broken ones.

- scissors
- twist tie or piece of wire about 10 cm long
- weight such as 1 of the following:
    - 1-inch bolt and several nuts (Many sizes may work.)
    - swivel or egg-type fishing sinker from ¾–1 ounce (or a combination of smaller sinkers)
    - pennies tied up in a piece of cloth
- 2 toothpicks
- (optional) 2 short pencils, pens, or pieces of chalk

For the Variation
Per class
- commercial come-back toy
- (optional) bell, buzzer, or whistle

For the Extension
Per group
- masking tape

52   Exploring Energy with **TOYS**

- meterstick
- roll-back toy constructed in the Procedure

SAFETY AND DISPOSAL

Cover sharp edges on the cans with electrical or masking tape. No special disposal procedures are required.

GETTING READY

1. Cut off the ends of all the cans and punch a hole in the center of each lid in advance. Cover sharp edges on the cans with electrical or masking tape.

2. Assemble one come-back toy as an example of the finished product. (See Instruction Sheet).

PROCEDURE

1. Demonstrate the motion of the commercial come-back toy by giving it a push. Ask some questions to review the concepts involved, such as, "What kind of energy does the toy have while it is moving? Where does that energy come from? Where does the energy go when the toy slows down and stops? Why does the toy return?"

2. Lead the students to speculate about what is inside the toy that stores energy.

3. After students propose some ideas about what stores the energy, ask them how they could find out which one is correct. Hopefully someone will recommend taking the toy apart. If your commercial come-back toy can be taken apart, open up the toy and let the students come up in small groups to see it, or pass it around. If not, explain that while this particular toy cannot be taken apart, the students are going to make a toy that has a similar mechanism.

4. Have the students construct homemade come-back toys either individually or in small groups by following the Come-Back Toy Assembly Instructions (provided).

   Troubleshooting toys that don't work is an important part of the students' learning process. It helps them to understand what features in the design are essential to making the toy work and is a good problem-solving exercise. In our experience about 50% of the toys will work immediately. Most of the rest will work after small adjustments involving tightening the rubber band or changing the amount or position of the weight. Occasionally, one will not work even after significant efforts to fix it. In this case, it is best just to start over with new materials.

5. Have the students write explanations of how their homemade come-back toys work.

EXPLANATION

➤ *The following explanation is intended for the teacher's information. Modify the explanation for students as required.*

When you do work by pushing the can, you give the can kinetic energy. Because the weight does not turn with the can, the rubber band winds up. Rubber bands store energy when they are stretched or twisted and release it later. The energy stored in the twisted rubber band is called potential energy. This potential energy is released and returns to kinetic energy as the rubber band unwinds, rolling the can back to you. The can continues to roll back even after the rubber band is completely unwound. This continued movement is caused by the can's inertia. This motion winds the rubber band in the opposite direction, again storing potential energy. When the rubber band is wound as far as it will go, the can stops and rolls back in the direction it came from. This process continues until friction converts all the kinetic energy into thermal energy, and the can stops.

VARIATION

A commercial come-back toy (with a hidden mechanism) is a good tool to use for convergent thinking. To use it in this way do the following:
1) Demonstrate the toy several times. Pet it and/or whisper to it to add a little drama. You can also whistle to call it back. 2) Have the students ask simple "yes" or "no" questions, one at a time, about the toy's construction. 3) Use a bell or buzzer to indicate questions that correctly describe the construction of the toy. (If you don't have a bell or buzzer, you can simply answer "yes.") Do not respond to questions to which the answer is no, or complex questions to which the answer would be partly yes and partly no. Students must listen and build on each other's questions to reach a solution. 4) When a student believes that he or she can describe the construction from beginning to end, give that student an opportunity to share his or her hypothesis with the class.

EXTENSION

Put a long strip of masking tape on the floor. Mark a starting point near the middle of the tape. Place the homemade can at the starting point and push it along the tape. As the can rolls back and forth along the strip, mark each point at which it turns around. Measure from the starting point to each turning point. Graph these distances, taking right of start to be positive and left of start to be negative or vice versa. (See Figure 1.) Discuss with the students the conclusion which can be drawn from the graph: the toy is continually losing energy. On each pass, the distance traveled is less than that of the previous pass, so the rubber band is wound less tightly. Thus, less energy is stored. The students may also have noticed that the toy was moving more slowly as it crossed the starting point each time, indicating a decrease in kinetic energy. This activity gives no information about where this "lost" energy is going. Later the students will learn that friction is converting it into thermal energy of the toy and the floor.

Figure 1: Turn-Around Distances

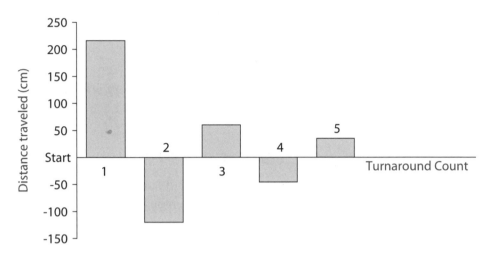

CROSS-CURRICULAR INTEGRATION

Art:
• Have students decorate their come-back toys. Students should wrap paper around their cans before assembling the toys.

Language arts:
• Have students do "I Search" papers. Tell students, "Think of a question about rubber bands or rubber that you would like answered. Research your questions using unconventional methods." Some examples are personal interviews, writing letters, e-mail, Internet, and phone calls.

Life science:
• Discuss the way the body stores chemical energy and uses it later to move muscles.

Math:
• Have students measure the distance the toy rolls forward and then backward. Find the ratio of the two distances. Ask students, "Is this ratio always the same?" Answering the question will require taking more data. Have the students look for patterns in the data rather than simply saying, "No, it is not the same." Is the ratio always the same for a particular toy but different for other toys? Does it get smaller or larger as the toy is used more? Does how hard you push it change the ratio? This activity could provide practice in long division, fractions, or percentages.

The Toy That Returns

55

FURTHER READING

Allison, L.; Katz, D. *Gee Wiz! How to Mix Art and Science or the Art of Thinking Scientifically;* Little, Brown: Boston, MA, 1983. (Students)

Faughn, J.; Turk, J.; Turk, A. *Physical Science;* Saunders: Philadelphia, PA, 1991. (Teachers)

Kirkpatrick, L.; Wheeler, G. *Physics: A World View,* 2nd ed.; Saunders: Philadelphia, PA, 1995. (Teachers)

CONTRIBUTOR

Mark Beck, Indian Meadows Primary School, Ft. Wayne, IN; Teaching Science with TOYS peer mentor.

HANDOUT MASTER

A master for the following handout is provided:
• Come-Back Toy Assembly Instructions
Copy as needed for classroom use.

The Toy That Returns
Come-Back Toy Assembly Instructions

1. Thread the rubber band through the holes in the lids. Insert a toothpick through each loop of the rubber band to keep it from pulling back through.

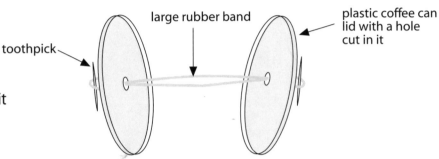

large rubber band

plastic coffee can lid with a hole cut in it

toothpick

2. Tie the twist tie or wire to the weight. Then tie it firmly to the center of one section of the rubber band. The twist tie or wire should be short enough so that the weight will not touch the side of the can, but some space should still remain between the weight and the rubber band. Crimp the twist tie or wire tightly so the weight does not slide along the rubber band.

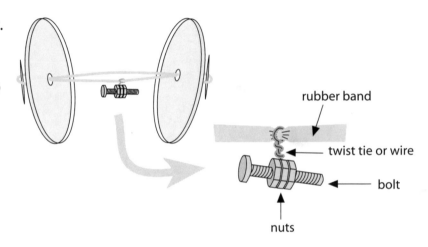

rubber band

twist tie or wire

bolt

nuts

3. Pull one lid through the can and snap on both lids.

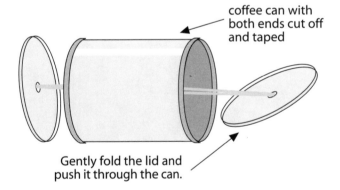

coffee can with both ends cut off and taped

Gently fold the lid and push it through the can.

4. Try rolling your can. If it does not come back, add more weight, adjust the position of the weight, or tighten the rubber band. To tighten the rubber band, you can wrap it a few more times around the toothpicks or replace the toothpicks with short pencils, pens, or pieces of chalk.

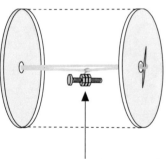

Make sure that the rubber band is tight enough and the twist tie or wire short enough to keep the weight from touching the side of the can.

☐ **✔ Concept Introduction**

☐ **Concept Application**

☐ **Synthesis**

How Much Energy?

...Students use spring-powered vehicles to explore the effects of varying the amount of kinetic energy.

✔ Time Required

Setup   negligible
Performance   40 minutes
Cleanup   5 minutes

✔ Key Science Topics

- energy conversion
- friction
- kinetic energy
- potential energy

✔ Student Background

Students should have been introduced to the concepts of work and energy conversion. They should also know that a stationary object, such as a book, moves in the direction of the force when a force acts on it. Students should be able to measure distances and plot points on a graph.

A Press-n-Roll truck

✔ National Science Education Standards

Science as Inquiry Standards:

- Abilities Necessary to Do Scientific Inquiry

  Students measure distance traveled, use graphs to present their data, and draw conclusions from the graphs.

  Students determine the relationship between number of presses or winds and distance traveled.

Physical Science Standards:

- Transfer of Energy

  Energy is a property of an object, and the quantity of energy in an object can be varied.

MATERIALS

For the Procedure
Per class
- meterstick or measuring tape
- Press-n-Roll truck by Geoffrey Inc., See and Go's Toot-Toot Loco by Blue Box®, or See Thru Racer by Small World Toys

The latter two toys come in trucks, locomotives, helicopters, etc. Any of them will work.

For the Extensions
❶ All materials listed for the Procedure, plus the following:
Per class
- stopwatch
- (optional) calculator

SAFETY AND DISPOSAL

No special safety or disposal procedures are required.

INTRODUCING THE ACTIVITY

Demonstrate the vehicle several times, storing a different amount of energy each time. Point out that the vehicle goes farther with more presses or turns. If you have done the activity "What Makes It Go?" ask students how this toy is different from the Push-n-Go® in the way it operates.

PROCEDURE

1. Ask students why the vehicle goes farther with more presses or turns. (Their answers should be related to the larger input of energy. If necessary, lead the discussion to this conclusion.)

2. Have students test their toys on a hard, smooth floor and collect, record, and graph data according to one of the following methods:

    a. If using the Geoffrey Press-n-Roll, have students record the distance the vehicle travels beginning with one press and going up to seven presses and graph the distance versus the number of presses.

    b. If using the Toot-Toot Loco or See Thru Racer, have the students record the distance traveled for each full- or half-turn of the key and graph the distance versus the number of turns.

3. Have students think about their graphs. Ask, "Do the points form a straight line? What does this tell you? Does the line slope up or down? What would it mean if the line were curved instead of straight?"

EXPLANATION

The following explanation is intended for the teacher's information. Modify the explanation for students as required.

The Press-n-Roll truck operates on the same basic principles outlined in the activity "What Makes It Go?" The truck rolls farther with each additional press of the rider's head because more energy is being supplied to the spring with each press. This energy is stored in the spring and released when the vehicle is let go. The spring in the Press-n-Roll is not a linear spring as in the Push-n-Go but a torsional spring (the kind in a windup watch). A ratchet mechanism keeps the spring from unwinding as the rider rises so that you can keep winding with more presses. The spring does not unwind until the rider gets all the way to the top. The Toot-Toot Loco and See Thru Racer vehicles also store energy in a spring, but the spring is wound by turning the key. Nothing prevents unwinding once the key is released, so the wheels must be held until it is time for the vehicle to move.

The students' graph should be a fairly straight line with an upward slope. (See the sample graphs in Figures 1 and 2.) This pattern shows that the number of presses or turns directly determines how far the vehicle will travel—the more presses or turns, the farther the vehicle will go. (If the students' graph sloped down instead of up, that would mean that the vehicle traveled less far with each additional press, which is not the case.)

Figure 1: Distance Traveled Versus Number of Presses in a Press-n-Roll Truck

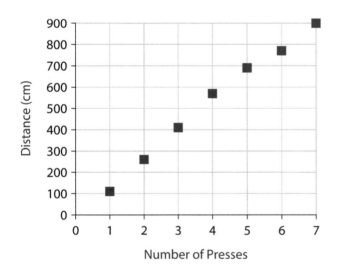

Number of Presses

Figure 2: Distance Traveled Versus Number of Turns in a See Thru Racer

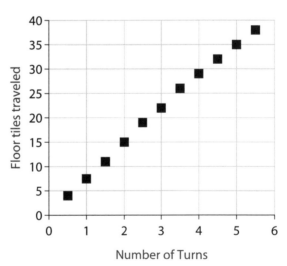

Number of Turns

The linear relationship between presses or turns and distance results from the fact that each press of the head or turn of the key stores the same amount of energy. (If your class also does the activity "Exploring Energy with an Explorer Gun," you may want to contrast the toy vehicles with the gun, in which the amount of energy stored per turn is not constant.) If students were to test several numbers of presses beyond seven and measure the distance, the graph would flatten out at about ten presses because the spring would be wound to its tightest position. Additional presses would not add more energy to the Press-n-Roll truck. This behavior cannot be observed with the Toot-Toot Loco or See Thru Racer because the key does not turn after the spring is fully wound.

The energy from the hand that presses down on the rider's head or turns the key is stored in the spring as potential energy, then converted into kinetic energy (KE). Greater initial KE means greater initial velocity, as shown in Figure 3. The distance traveled is determined by the initial kinetic energy and the size of the friction force (which is constant in this experiment). The vehicle loses kinetic energy as this energy is changed into thermal energy in the process of overcoming friction. So the greater the initial kinetic energy, the greater the distance that is traveled before all of the kinetic energy is changed to thermal energy. If we lower the friction force, the vehicle would travel farther before all the kinetic energy is converted into thermal energy.

If the graph were curved instead of straight, that would mean that the number of presses and distance traveled were not directly proportional— for example, six presses would not travel twice as far as three presses, which is not true for these toys.

EXTENSIONS

1. Call students' attention to the fact that the speed with which the vehicle is traveling at the beginning of the motion varies. (Visually compare two presses or turns to five presses or turns, for instance.) Have students measure the time required to travel the first 1-m length for different numbers of presses or turns. Graph the results and talk about why the graph has the shape it does. (See Figure 4. The time decreases with increasing numbers of presses or turns, indicating the vehicle is traveling faster. With the Press-n-Roll, after about 10 presses, the time doesn't change with more presses, indicating that the spring is fully wound.) Students could also calculate the toy's average speed for the first meter by using the formula speed equals distance divided by time. (See Figure 3.)

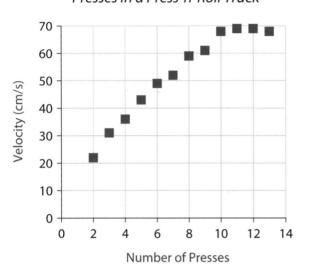

Figure 3: Velocity Versus Number of Presses in a Press-n-Roll Truck

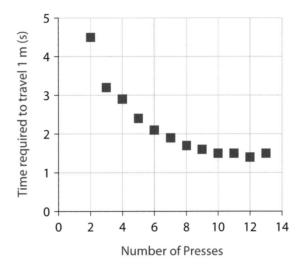

Figure 4: Time to Travel 1 m Versus Number of Presses in a Press-n-Roll Truck

2. Repeat the experiment on a carpeted surface. A linear relationship will still exist between presses and distance traveled, but all of the distances will be smaller due to the larger friction force.

CROSS-CURRICULAR INTEGRATION

Math:

• Have the students measure time and distance traveled by the Press-n-Roll truck and practice graphing and charting.

FURTHER READING

Ardley, N. *The Science Book of Energy;* Harcourt, Brace, Jovanovich: Orlando, FL, 1992. (Students)

Kirkpatrick, L.D.; Wheeler, G.F. *Physics: A World View,* 2nd ed.; Saunders: Philadelphia, PA, 1995. (Teachers)

CONTRIBUTORS

Ryan Corris, Hopewell Elementary School, West Chester, OH; Teaching Science with TOYS Research Teacher, 1996.

Hope Spangler, Graduate Assistant, Department of Physics, Miami University, Oxford, OH, 1991–93.

Exploring Energy with an Explorer Gun

...Students observe how an Explorer Gun stores energy and works.

✔ Time Required

Setup	10	minutes
Performance	60	minutes
Cleanup	5	minutes

✔ Key Science Topics

- averages and ranges in data
- kinetic and potential energy
- work

✔ Student Background

Students should be familiar with the concepts of work, potential energy, and kinetic energy. The activity "What Makes It Go?" provides a good introduction to these concepts. Investigating the Explorer Gun® will extend students' understanding of these concepts. Although not essential, some previous experience with projectile motion would be helpful in understanding this activity, particularly if the Variation or the Extension is used. This experience could be provided by doing the *Teaching Physics with TOYS* activity "Two-Dimensional Motion."

An Explorer Gun with disk "ammunition"

✔ National Science Education Standards

Science as Inquiry Standards:

- Abilities Necessary to Do Scientific Inquiry
  Students control variables during the experiment.
  Students collect and record data and make a graph to assist in interpreting the data.
  Students use their data to predict what will happen in a new situation.

Physical Science Standards:

- Transfer of Energy
  Energy can be stored in a spring and then transformed into kinetic energy of other parts of the object.

MATERIALS

For the Procedure
Per group
- Explorer Gun by Park Plastic

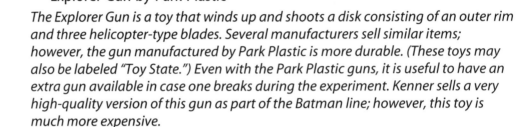

The Explorer Gun is a toy that winds up and shoots a disk consisting of an outer rim and three helicopter-type blades. Several manufacturers sell similar items; however, the gun manufactured by Park Plastic is more durable. (These toys may also be labeled "Toy State.") Even with the Park Plastic guns, it is useful to have an extra gun available in case one breaks during the experiment. Kenner sells a very high-quality version of this gun as part of the Batman line; however, this toy is much more expensive.

- marker or paint
- meterstick, metric measuring tape, or trundle wheel
- (optional) meterstick, desk, or table

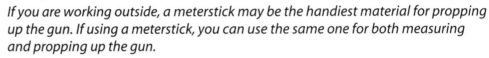

If you are working outside, a meterstick may be the handiest material for propping up the gun. If using a meterstick, you can use the same one for both measuring and propping up the gun.

For Class Discussion
Per class
- a toy that uses a twisted (not compressed) spring, such as a See-Thru-Loco, Crazy Wheel, or Press-N-Roll (by Lil Hands)

For the Extension
Per group
- paper
- scissors
- tape
- clay
- sheets of Styrofoam™ or similar lightweight material

SAFETY AND DISPOSAL

Remind students not to fire the Explorer Gun at anyone. No special disposal procedures are required.

GETTING READY

1. Trial test each Explorer Gun to determine the maximum number of half-turns the gun will allow. This number varies from one gun to another but will be about six or seven. Measure the distance the disk travels with the maximum number of winds, and make sure the area to be used for the experiment is large enough to accommodate this distance.

2. The Procedure of this activity does not present step-by-step instructions for the experiment, because the exact design is left up to you. As long as the number of half-turns of the disk is the variable and all other factors are held constant, a variety of experimental designs will yield good results. Ideally, students should test at least six different numbers of half-turns.

This procedure is written as if the experiment were designed by you, the teacher. If your students have had sufficient experience with scientific experiments, you could allow each group to propose an experiment design. The class should then critique all the designs and choose one for the whole class to follow. If necessary, you can guide them in how to correct flaws in the design.

Since the Explorer Gun is a toy rather than a precision scientific instrument, repeated measurements with the same gun will not give identical results. Thus, it is ideal for each group to do several trials for all six numbers of half-turns and average their results. However, this may be too time-consuming. Alternatively, you could have each group test each number of half-turns once and then collect class data for averaging. (See the Explanation for further discussion.) Whether the groups do each number of half-turns once or several times, they should all do the same six numbers of half-turns.

PROCEDURE

Part A: Discussing the Experiment

1. Show the students the Explorer Gun and demonstrate the proper method for firing it. Fire it several times, winding it a different number of half-turns each time. Explain to the students that a scientist might wish to investigate the relationship between the amount of winding and the distance flown. Announce that the purpose of the activity is to discover whether the disk will fly twice as far if it is wound twice as many times. Write the question, "Does winding the gun six half-turns cause the disk to travel twice as far as three half-turns?" on the board, or have a student do so. (If the students have already done "How Much Energy?" an alternative would be to tell them they want to determine whether the Explorer Gun behaves in the same way as the Press-n-Roll.)

Explain the following ideas: "An experiment must be set up so that it can be exactly repeated numerous times. Part of making this experiment repeatable is keeping the gun consistently level and always firing from the same height." A height of 1 m is convenient since the top of a vertical meterstick can be used as a reference. If you have access to movable tables or desks that are all the same height, you might have the students place the gun against a table edge with the disk sticking out in front of the table as a guide for judging whether the gun is level.

2. Describe the experimental design you would like your students to use. (See Getting Ready.)

3. Have the students name any conditions of the experiment that are being kept constant and any variables that are being tested. List these on the board.

Part B: Conducting the Experiment

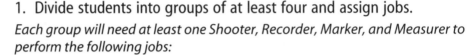

1. Divide students into groups of at least four and assign jobs.

Each group will need at least one Shooter, Recorder, Marker, and Measurer to perform the following jobs:
- *Shooter—to turn the disk and fire the gun,*
- *Recorder—to help make sure the gun is level and record the data,*
- *Marker—to spot and mark the landing point, and*
- *Measurer—to measure the distance to the landing point.*

2. Demonstrate the method that you want students to use for counting half-turns used to wind the gun. One method is to count half-turns by placing the thumb on top of the disk and then turning until the thumb is on the bottom. Alternatively, you could count the number of "clicks" heard while winding. Or you could put a dot of bright marker or paint on the disk and count the number of times the dot travels halfway around the circle. Be sure to warn the students not to force the gun beyond the maximum number of half-turns or it will break.

3. Move to the gym or outdoors and distribute materials.

4. Have students carry out experiments and collect data.

Part C: Compiling Data

1. Upon returning to the classroom, have each group use its data to make a line or bar graph of the distance traveled in meters versus the number of half-turns. If each group did several trials for each number of half-turns, each group should average its own results before making the graph. If students are making line graphs, tell them not to connect the dots. (See Figure 1.)

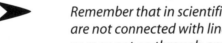

Remember that in scientific line graphs, points are plotted, but the individual dots are not connected with lines, although a best-fit curve may be drawn (which may or may not go through any of the dots). If you wish, your students may draw a best-fit curve by eyeballing the trend.

Figure 1: Effect of Winding on Projectile Range of Explorer Gun

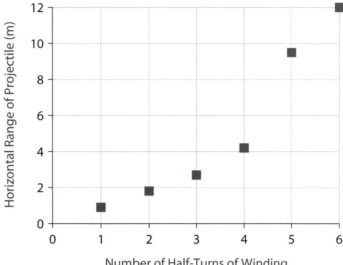

Number of Half-Turns of Winding

2. Have one student from each group put the group's data or graph on the board. (If one group's data stand out as quite different from the rest, have them demonstrate their procedure. Compare their procedure to the class list of factors that should be kept constant.) Students should then average all the values from all groups for a given number of half-turns.

3. (optional) A graph using these average values can be drawn on the board. Although this step is optional, it provides an excellent platform for introducing or reviewing measurement uncertainties. (See the Explanation.)

CLASS DISCUSSION

Lead the students through a discussion of how energy is stored by the gun and then converted into kinetic energy when the trigger is pulled. Since students may not be familiar with springs that are twisted rather than compressed, it is useful to have such a spring to show the students. One source is a See-Thru-Loco that operates on the same principle and has a clear plastic body, so the spring and gears can be seen. Another is a Crazy Wheel. The tire is easily removed from the wheel to reveal the spring.

You should also make the connection between the initial velocity of the disk and the distance it travels before hitting the ground. This discussion should be a review of ideas from previous studies of motion.

Return to the question that initiated the experiment: "Does winding the gun six half-turns cause the disk to travel twice as far as three half-turns?" The points on the graph will not fall in a straight line, and the answer is clearly no.

To illustrate that the first few half-turns do not store much energy, compare the increase in distance traveled in going from one to two half-turns with the increase from five to six half-turns. The students should recall that when winding the gun they had to exert a larger force near the end than at the beginning. Make sure the students understand the connection between the size of the force and the amount of energy stored.

If you did a graph with the class averages, you will also want to discuss why this is better than using just one trial. Use the graph to predict the distance traveled for some number of turns that you did not measure (for example, 2¼ complete turns). Have each group measure the distance for this amount of input energy or have one group make this measurement at a later time and report back to the class. (Ideally, they should measure several trials and average the results.) This illustrates the usefulness of the graph and ties back into the discussion of the meaning of average as a "best estimate." Discuss the fact that the usefulness of the average as a predictor depends upon the range of the data—the wider the range, the less reliable the average as a predictor. (See Explanation.)

EXPLANATION

The following explanation is intended for the teacher's information. Modify the explanation for students as required.

As you wind the Explorer Gun, you do work. Energy from your body is stored in the spring as potential energy and then later converted into the kinetic energy of the disk.

You do work as you wind the gun because you apply a force to the disk and the point where the force is applied moves as a result. The work required for the first turn of the disk is not the same as that required for the fourth because more force is required with each subsequent turn. When you turn the disk, a spring is twisted, as in watches that use mechanical energy rather than batteries. The spring stores the energy. The amount of work done, and thus the amount of energy stored, depends on the total number of half-turns. The relationship between the number of half-turns and the amount of energy stored is not linear because the force required for each subsequent turn is not constant.

When you pull the trigger of the gun, the spring untwists, turning the axle to which the disk is attached and pushing the disk forward off the axle. The amount of stored energy in the spring (that is turned into kinetic energy) determines the velocity of the disk when it leaves the gun. Although it is not necessary to raise the issue with the students, you should understand that this is a more complicated case than, for example, a dart gun that stores energy by compressing a spring. Part of the energy stored in the Explorer Gun goes into kinetic energy associated with the rotation of the disk as well as into the kinetic energy associated with the forward motion, as in the dart gun.

The velocity of the disk as it leaves the gun determines the distance it travels before hitting the ground. As soon as the disk leaves the gun, it begins to fall due to the force of gravity. The amount of time it takes to drop is a function only of the height from which it begins, which is the same for all trials. The distance the disk travels before hitting the ground is the product of this amount of time in the air and the initial velocity. If the disk is launched at an upward angle (up to 45°), the time in the air will be increased, and as a result the distance traveled by the projectile will be increased.

Although the number of half-turns is supposed to be the only variable in the experiment, the springs in the various guns are not identical and may be a hidden variable (an unintentional difference in experimental conditions that may affect results), depending on how you compile the data. For example, if each group does several trials for each number of half-turns, and averages and graphs ONLY their own data, the spring is a constant for that group of data because all the data are generated using the same gun. But if each group collects only one value for each number of half-turns and all groups' data are used to create a class average, the spring becomes a hidden variable, because the data used to create the class average are generated using different guns with different springs. If all the groups combine their own averages into a class average, the spring will again be a hidden variable for the same reason as before. (This method of combining data will still provide satisfactory data for drawing conclusions; however, the range of values will be increased.)

Good experiments generate data that allow experimenters to make predictions. The average values generated from many trials are a "best estimate" prediction of the data that would be generated if the experiment were repeated. The certainty of this "best estimate" depends, in part, on the number of values collected, the range (span between smallest and largest), and how closely these values are clustered near the average. If we have very few values and they have a very wide range, we might question the usefulness of the average of those data as a predictor. If we have many values that still have a wide range, but many are clustered near the average value, we may feel more confident in the average of those data as a predictor. If we have many values, they have a narrower range, and many are clustered near the average, we may feel even more confident in this average as a predictor. (See Figure 2.)

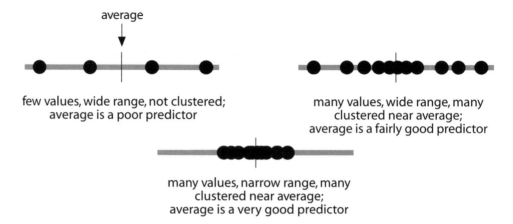

few values, wide range, not clustered;
average is a poor predictor

many values, wide range, many
clustered near average;
average is a fairly good predictor

many values, narrow range, many
clustered near average;
average is a very good predictor

Figure 2: The average of a set of values is a more reliable predictor when many values are clustered in a narrow range near the average.

Hidden variables (such as wind or the spring) can make data more scattered and less reliable. Usually, the more data you collect, the less impact a hidden variable like the spring will have on the reliability of your averages. If each group did several trials for each number of half-turns, and therefore you have many more values when computing your averages, these averages should be more reliable predictors of the values you would get if you repeated the experiment. The range in distance values for a given number of half-turns and how many of these values are fairly close to the average will give you an idea about the reliability of your averages as predictors. The smaller the range of distance values, the closer you would expect the new measurement to be to the average. Your students can test the reliability of one of the class averages by repeating part of the experiment using a randomly chosen gun and then comparing the new result to the average.

VARIATION

Instead of the activity described above, a more general investigation could be done. Ask the class, "How can we get the disk to go the greatest possible distance?" If student answers do not include manipulating the firing angle, height above ground, and number of winds, then lead them to those ideas. While students may have additional ideas, such as making the disk lighter, they will probably be difficult to test.

Divide the class into three or more groups: one to do the winding experiment, one to investigate the effect of the firing angle, one to investigate the effect of height above the ground, and other groups if necessary to test other variables students have proposed. Caution the students to keep everything constant except the one variable they are investigating.

The data can be presented in tables or graphs. The follow-up discussion should include how to combine all three (or more) variables to maximize the distance the disk travels. For further discussion of the effects of the three variables mentioned, see the *Teaching Physics with TOYS* activity "Two-Dimensional Motion."

EXTENSION

Pose the question, "How could we change the design of the gun or disk to make the disk go farther (or not as far)?" Brainstorming will probably produce several ideas that can be tested and several that cannot. Possibilities that cannot easily be tried are to use a spring that winds more easily, remove the blades from the disk, and make it easier to pull the trigger.

Some ideas that can be checked by experimentation are making the disk heavier, using a solid disk, and increasing the diameter of the disk. Groups should first make a measurement using the original disk to compare to the results after the changes are made. Various amounts of clay can then be added to the rim of the disk and distributed evenly so the disk is still balanced. Paper can be taped over the front of the disk to eliminate the effect of the blades and simulate a solid disk. If you have access to sheets of a lightweight material such as Styrofoam, you could try increasing the diameter of the disk by taping a Styrofoam ring around the outside of the disk.

This extension has a number of important aspects. Foremost, the students are actively involved in "What will happen if…" thinking. They must decide how to test their ideas and carry out the experiment. Discussion of the results is also important. Encourage the students to speculate about why the results were as they were. Do not feel that you must be able to give a scientific explanation in each case. This discussion might be used as a springboard into further reading on a variety of topics, including aerodynamics and spinning objects.

ASSESSMENT

The following is a list of possible assessment questions. You might want to use some of them as group discussion questions, journal writing assignments, or part of a written examination.

1. When firing the Explorer Gun, why does the disk go so much farther with six half-turns than with two half-turns?

2. Would you expect the disk to go farther if the gun is straight when fired or if the gun is tilted up slightly? Why?

3. When does the Explorer Gun have the greatest potential energy?

4. Give an example of a change of energy from one form to another in the Explorer Gun.

5. Why is it good procedure to take several measurements and average them?

CROSS-CURRICULAR INTEGRATION

Art:
- Have students draw, paint, or make clay sculptures of objects that store energy. A group collage of energy-storing objects would be an interesting class project.
- Students may also investigate what is meant by the phrase "kinetic art," and find pictures of such artwork or even construct some.

Language arts:
- Once the students have been introduced to the topic of potential versus kinetic energy, they may begin to notice the wide variety of objects in the world that store energy and the variety of ways in which that energy is released. Conduct a discussion using students' observations about the work done by a plant growing up through a crack in the pavement, extending the crack further, and crumbling the concrete.
- Conduct a brainstorming activity in which students list as many energy-storing objects as possible as a springboard for a writing activity. Use the list to inspire creative writing in which some of these objects and their energy-storing properties are important.

Math:
- If your students have not yet been introduced to the technique of averaging, you may even wish to conduct this lesson as a math activity rather than a science activity, particularly if your science time is limited, in order to give the students a concrete example of averaging.

Social studies:
- Tie the discussion of uncertainties in measurements to opinion polls and their uncertainties. (For example, students may hear on television that 35% plus or minus 3% of the population favor a particular public policy. What does that "plus or minus" mean?)

FURTHER READING

Frances, N. *Super Flyers;* Addison-Wesley: Reading, MA, 1988. (Students)

Gartrell, J.E., Jr.; Schafer, L.E. *Evidence of Energy;* National Science Teachers Association: Washington, D.C., 1990. (Teachers)

Kirkpatrick, L.D.; Wheeler, G.F. *Physics: A World View,* 2nd ed.; Saunders: Philadelphia, PA, 1995. (Teachers)

Walpole, B. *Fun with Science: Movement;* Warwick: New York, NY, 1987. (Students)

Pop Can Speedster

...Students explore how elastic potential energy is stored in a twisted rubber band and converted to kinetic energy.

✔ **Time Required**

Setup	20 minutes
Performance	45 minutes
Cleanup	5 minutes

✔ **Key Science Topics**

- friction
- kinetic energy
- potential energy

✔ **Student Background**

Students should have been previously introduced to the concepts of potential and kinetic energy and friction.

A Pop Can Speedster

✔ **National Science Education Standards**

Science as Inquiry Standards:

- Abilities Necessary to Do Scientific Inquiry

  Students observe the behavior of the speedster as they change several variables one at a time.

  Students use their observations to predict what will happen in a new situation.

Physical Science Standards:

- Transfer of Energy

  Energy is a property of an object which is sometimes associated with its mechanical motion.

  Energy can be stored in the rubber band and transformed into kinetic energy through the interaction of the can and the floor.

MATERIALS

For Getting Ready only
Per class
- 8-penny nail
- hammer

For the Procedure
Per speedster
- 2 medium rubber bands (One should be about ¼ inch wide; the other can be narrower.)
- empty 12-ounce soft-drink can (rinsed and dried)
- 1 or more medium-sized craft beads with holes large enough for the smaller rubber band to slide through
- long pipe cleaner (at least 12 inches long)

Per group or per class
- materials to make a ramp, such as stiff cardboard or plywood propped against a book or box

For the Extensions
❸ Per class
- more rubber bands
- rubber cement
- salt
- sandpaper
- glue
- other materials to increase the traction of the can

SAFETY AND DISPOSAL

Caution students that the openings of the cans might have sharp edges. No special disposal procedures are required.

GETTING READY

With a hammer and nail, punch a hole in the center of the bottom of each rinsed beverage can. Assemble one pop can speedster according to the instructions in Part A of the Procedure. If the speedster does not roll fairly rapidly or simply spins in place, try a stronger or weaker rubber band on the inside.

INTRODUCING THE ACTIVITY

Demonstrate the pop can speedster to the class. Ask students how it is related to the ideas of kinetic and potential energy. Explain that they will be building their own speedsters and investigating their behavior.

PROCEDURE

Part A: Assemble the Speedsters

> *If students will make one speedster per group, you should demonstrate the complete procedure, then let each group make a speedster while you circulate to spot problems. If each student will make his or her own speedster, it is more efficient to have the whole group do each step simultaneously as you demonstrate.*

1. Stretch the wide rubber band around the outside of the can. It works best if it is placed about 1–2 inches from the bottom of the can and lies completely flat against the can.

2. Push the pipe cleaner through the hole in the bottom of the can and out the top. (See Figure 1.)

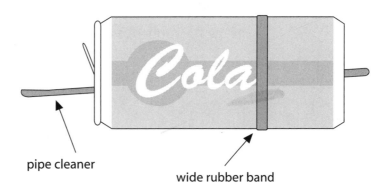

pipe cleaner

wide rubber band

Figure 1: Push the pipe cleaner through the can.

3. Make a hook in the end of the pipe cleaner at the top of the can. (See Figure 2.)

Figure 2: Make a hook in the end of the pipe cleaner at the top of the can.

4. Put the narrower rubber band on the pipe cleaner hook and press the hook down tightly or twist it a few times to hold the rubber band.

5. Hook the rubber band under the tab. (See Figure 3.)

Figure 3: Hook the rubber band under the tab.

6. Gently pull the pipe cleaner back through the hole in the bottom of the can, stretching the rubber band through the can.

7. Thread the craft bead onto the free end of the pipe cleaner and down to the rubber band.

More than one bead might be necessary depending on their size. The purpose of the bead is to keep the pipe cleaner from rubbing the bottom rim of the can.

8. Gently open the hook holding the rubber band and slide the rubber band to a position about 1½ inches from the center of the pipe cleaner.

9. Fold the pipe cleaner in half. Twist the two halves together above and below the rubber band to stiffen it. (See Figure 4.)

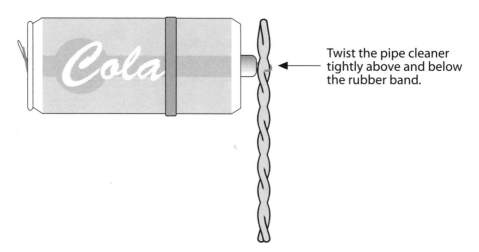

Twist the pipe cleaner tightly above and below the rubber band.

Figure 4: Fold the pipe cleaner in half and twist the halves together.

10. Wind up the speedster by holding the can and twirling the pipe cleaner around the can with your finger. Set the speedster on the floor and release the pipe cleaner. (If the can spins in place instead of rolling, the inner rubber band may be too strong.) If the can does not roll at all, try winding the pipe cleaner several more times. If the can rolls in a circle instead of rolling straight, change the location of the outer rubber band or the angle of the pipe cleaner relative to the can.

Part B: Investigate with the Speedsters

1. Have students investigate with the speedster (in groups, if desired) to answer the questions on the Data Sheet (provided).

2. When the students are finished with the questions, discuss some of the key points in the activity. Be sure to reinforce what the students discovered as well as provide closure to the activity.

EXPLANATION

> *The following explanation is intended for the teacher's information. Modify the explanation for students as required.*

When the speedster is wound up, energy is stored in the rubber band. The force your hand exerts while winding the rubber band does work, which is stored in the wound rubber band as elastic potential energy. When you wind up the speedster enough and let it go on the floor, it will move. The elastic potential energy in the rubber band is now being converted to kinetic energy of the speedster.

Changing how tightly the rubber band is wound changes both the initial speed and the distance the speedster travels. The distance changes because the total energy stored has changed. The initial speed changes because a tighter rubber band exerts more force on the can. The force that actually makes the can roll is the friction force between the can (and outer rubber band) and the floor. As the inner rubber band pushes on the can in its attempt to unwind, the can in turn pushes on the floor, and the floor pushes back (Newton's third law). When the rubber band is wound tighter, all these forces are increased until the maximum possible friction force is reached (determined by the roughness of the surfaces and the weight of the speedster). After this limit is reached, additional force will just cause the speedster to spin without rolling as the can slips against the floor. The purpose of the outer rubber band is to increase the size of the maximum friction force.

Question 4 on the Data Sheet begins to introduce students to the concept of gravitational potential energy, which you do not want to explain at this point. The question is just to get them thinking about this concept.

EXTENSIONS

1. Explore the relationship between the number of winds and the distance (or time) the speedster goes before stopping. If both distance and time are measured, average speed can be calculated.

2. Determine the maximum number of turns the pipe cleaner can be wound before the can spins instead of rolls.

3. Develop an experiment that explores the use of different traction methods. Methods could include using more rubber bands, coating the can with rubber cement and rolling it in salt, or gluing sandpaper to the outside of the can.

4. Have each student take a speedster home, think of a design change, and implement it. Students should bring their altered toys to school the next day and be prepared to demonstrate and explain how their changes affected the toys' motion.

CROSS-CURRICULAR INTEGRATION

Language arts:
- Read aloud or suggest that students read the following book:
  - *Who Can Fix It?,* by Leslie Ann MacKeen (Landmark Edition, ISBN 0933849192)
  When Jeremiah's car breaks down on the way to his mother's house, several animals stop by to offer amusing solutions to his problem. Compare fixing the car in the story to the problem-solving some students had to do when their speedsters did not work properly.

Life science:
- Study how the human body stores energy for later use.

Math
- If Extension 1 is done, have the class practice averaging and graphing.

Social studies:
- Study energy production and storage mechanisms such as batteries, passive solar energy systems, and pumped storage at hydroelectric plants.
- Create a timeline showing how vehicles have been powered over time.
- Research the development of different energy sources that power transportation. Examples are gasoline engines, diesel engines, steam engines, and electric motors.

FURTHER READING

Ardley, N. *The Science Book of Energy;* Harcourt Brace Jovanovich: Orlando, FL, 1992. (Students)

Eichelberger, B.; Larson, C. *Constructions for Children: Projects in Design Technology;* Dale Seymour: Palo Alto, CA, 1993. (Students)

Milson, J. "Tin Can Racer Derby," *Science and Children.* 1986, *24*(2), 17–19. (Teachers)

Kirkpatrick, L.D.; Wheeler, G.F. *Physics: A World View,* 2nd ed.; Saunders: Philadelphia, PA, 1995. (Teachers)

CONTRIBUTORS

Glen Gillen, Graduate Student, Physics Department, Miami University, Oxford, OH; 1994.

Kathleen Hayse, Worth County Schools, Grant City, MO; Teaching Science with TOYS, 1993.

Susan Higgins, Union High School, Biggsville, IL; Teaching Science with TOYS, 1995.

Kris Holder, Carroll Elementary School, Carroll, OH; Teaching Science with TOYS, 1996–97.

HANDOUT MASTER

A master for the following handout is provided:

- Data Sheet

Copy as needed for classroom use.

Name _____   Date _____

Pop Can Speedster
Data Sheet

1. Does the number of winds of the rubber band affect the toy's speed? Try it.

    _____

2. What is the function of the rubber band on the outside of the can? Take it off and test the speedster. What happens? Were you right?

    _____

    _____

    _____

3. What do you think will happen if the pipe cleaner is not allowed to touch the ground when the speedster is released? Try it. What happened? Can you explain why this happened?

    _____

    _____

    _____

4. Predict what will happen if you let the speedster climb a ramp. Explain why you think so. Try it. Describe what happens.

    _____

    _____

    _____

    _____

5. Use energy ideas to explain what happens when the speedster is wound up and let go.

    _____

    _____

    _____

    _____

 Reproducible page from *Exploring Energy with **TOYS*** published by Terrific Science Press™

Ladybug, Ladybug, Roll Away

...Students build their own rolling toys and investigate energy transformations.

✔ Time Required

Setup   5–45 minutes (depending on which
                      design is used)
Performance   80   minutes (two 40-minute periods)
Cleanup   5   minutes

✔ Key Science Topics

- energy conversion
- kinetic energy
- potential energy
- work

A homemade ladybug toy

✔ Student Background

Students should be familiar with the concepts
of force, work, and potential and kinetic
energy.

✔ National Science Education Standards

Science as Inquiry Standards:

- Abilities Necessary to Do Scientific Inquiry
  Students base predictions on the evidence they have gathered.
  Students form a logical argument about cause-and-effect relationships.

Physical Science Standards:

- Transfer of Energy
  Energy stored in rubber bands can cause the ladybug toy to move.

✔ Additional Process Skills

- measuring   Students measure the distance the toy moves when the string is
                     pulled different amounts.

- communicating   Students communicate their understanding of energy transformations
                     through a written summary.

MATERIALS

For Getting Ready
- 1 of the following:
  - ○ nail, candle, and lighter with pliers or hot pad
  - ○ circle-drawing compass
- 3-inch x 5-inch index card
- scissors
- rubber band
- ruler
- (optional) single-hole punch
- (optional) small hand saw
- (optional) electric drill and drill bit

For Introducing the Activity
- several toys that move as a result of being pushed, such as a small toy car, ball, or marble
- index card with rubber band prepared in Getting Ready
- (optional) pull-back car or commercial version of the ladybug toy

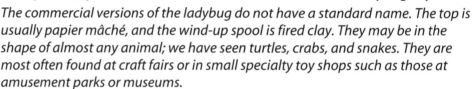

The commercial versions of the ladybug do not have a standard name. The top is usually papier mâché, and the wind-up spool is fired clay. They may be in the shape of almost any animal; we have seen turtles, crabs, and snakes. They are most often found at craft fairs or in small specialty toy shops such as those at amusement parks or museums.

For the Procedure
Per student or group
- disposable plastic bowl (Use red for a ladybug. Solo® brand works well.)
- 2 medium-sized, wide (up to 1 cm) rubber bands (Rubber bands from vegetable bunches work well.)
- 1 of the following sets of materials:
  - ○ full (or nearly full) plastic spool of thread with honeycomb center and a long rubber band (about 36 cm long when cut)

If you are using thread spools, send each student home with a request form 4–5 weeks ahead of time.

  - ○ 5-cm section of 1-inch dowel rod and 2 small rubber bands

You will probably need to purchase a longer dowel rod and cut it into 5-cm pieces. (See Getting Ready.)

  - ○ film canister with lid; enough popcorn kernels to fill the canister; extra medium-sized, wide rubber bands; and 1 long rubber band (about 36 cm long when cut)

Photo developing shops are often willing to save film canisters for teachers.

- 2 metal paper fasteners
- approximately 1 m embroidery floss, kite string, or heavy-duty thread
- small plastic ring or button

- tape
- meterstick or measuring tape
- (optional) for ladybug decorations: black pipe cleaners, felt or fabric remnants

This toy could be decorated in many other ways. Our teachers have made frogs, turtles, and snowmen. Use your imagination to correlate toy designs with the season or a story the students have read.

For the Variation

All materials listed for the Procedure plus the following:

Per student or group

- small balloon
- wax paper or plastic wrap
- papier mâché
- knife

For the Extension

Per class

- wide variety of types and sizes of bowls, rollers, rubber bands, and materials for decoration.

GETTING READY

For Introducing the Activity

Fold a 3-inch x 5-inch index card in half vertically. On the 3-inch edges, cut two slits about an inch apart and an inch deep. Open the card partway. Slip a rubber band through the four slits. (See Figure 1.)

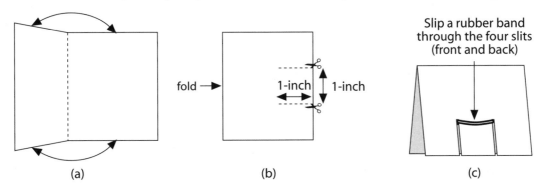

Figure 1: Prepare the 3-inch x 5-inch card. (a) Fold the card in half. (b) Make four 1-inch slits by cutting two slits on the non-folded edges of the card. (c) Slip a rubber band between the four slits.

For the Procedure

Build one ladybug ahead of time (see Ladybug Assembly Instructions, provided) and make sure the rubber bands are the correct size. (The film canister or spool should be suspended with a minimal amount of sagging.)

Make holes in all of the bowls and dowel rods or film canisters to be used by students, as described below.

Make two small holes on opposite sides of the bowl and one hole about 2 cm behind the center of the bowl to provide the location for the metal fasteners and thread to be attached to the toy. (See Figure 2.) This is easily accomplished by heating a small nail and placing it in the desired location on the bowl. (Heat the nail by holding the head of the nail with pliers or a hot pad and putting the sharp end in the flame of a candle.) This melts a small hole in the bowl. Holes can also be punched with a sharp object such as a single-hole punch (side holes only) or a circle-drawing compass, but sometimes the bowls crack, so have some extra bowls on hand.

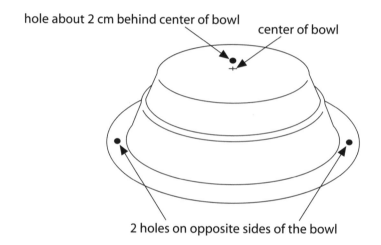

hole about 2 cm behind center of bowl   center of bowl

2 holes on opposite sides of the bowl

Figure 2: Make holes in the bowl.

If you are using a dowel rod, cut it into sections approximately 5 cm long. Drill a hole through each piece of dowel rod at each end about 1 cm from the ends. The hole must be large enough to thread a rubber band through. (See Figure 3.)

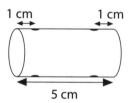

1 cm   1 cm

5 cm

*Figure 3: Drill holes 1 cm from the ends
of each section of the dowel rod.*

If you are using film canisters, make two holes about 1 cm apart in the lid and in the bottom of the can. A heated nail or small hand drill works well for this. The holes must be big enough for the long rubber band to go through.

Before copying the Ladybug Assembly Instructions, mark out the sections of step 2 that do not apply to the materials you are using.

INTRODUCING THE ACTIVITY

1. Before distributing materials and worksheets to students, display several different types of toys (such as a toy car, ball, or marble) that require an applied force to make them move. Remind students that the toys will remain at rest until a force is applied to them. The toys will then move in the same direction as the applied force. Demonstrate and discuss.

➤ *These concepts are embodied in Newton's first and second laws. However, there is no need to refer to the laws by name unless the students have recently studied them.*

2. Press the 3-inch x 5-inch card (see Getting Ready) open on the table, stretching the rubber band. Point out that you have exerted a force but the card isn't moving. Now let go and watch the card pop up. It does not move in the direction you pushed. Discuss the observations in terms of potential and kinetic energy.

   Alternatively, you could use a stored energy toy such as a pull-back car or a commercial version of the ladybug toy to start the same discussion.

3. Explain to students that they will be making a stored energy toy today.

PROCEDURE

Part A: Making the Ladybugs

Pass out materials and Ladybug Assembly Instructions. Have students follow the instructions to assemble the ladybug. (Alternatively, you could just give the instructions orally one step at a time.) After the toys are completed, collect and put them away until you are ready to move on to the science activity, using the Data Sheet (provided).

Part B: Investigating the Motion of the Ladybugs

1. Redistribute ladybugs for use during the science activity.

➤ *It may be necessary to adjust the tension of the string on the toy at this point. Have students check to make sure that the string is wound tightly around the roller before continuing with the activity. If it is not, remove the rubber bands from the fasteners, tighten up the string, and reattach the rubber bands to the fasteners.*

2. Have students experiment with their toys and complete the Data Sheets in small groups, even if each student made a toy.

➤ *The measurements are much easier to accomplish with several students working together, and the students can choose to make measurements with the toy that moves most nearly in a straight line. Also, students learn more when they can discuss the questions before writing their answers.*

3. Discuss the results with the students, reviewing the important energy concepts involved.

EXPLANATION

The following explanation is intended for the teacher's information. Modify the explanation for students as required.

The operation of this toy can be explained from a forces point of view or an energy point of view. We want to concentrate primarily on energy, but let's begin by thinking about forces. Newton's third law of motion says that if Object A exerts a force on Object B, then Object B exerts a force of the same size but opposite direction on Object A. In other words, if you push on the table, then the table is pushing back on you just as hard. Newton's third law describes forces that are acting simultaneously. As the roller in the toy turns, the friction rubber bands push backward on the floor, and the floor pushes forward with an equal force on the rubber bands. This force from the floor moves the toy.

When you pull the string on the toy, energy is transferred from you to the toy. You have done work on the toy. Remember that the definition of work is force times the distance the force moved (assuming the force and distance are in the same direction). In this case it is very clear that you exerted a force and the point of application of the force (the button or ring) moved. Work is also done on the toy car or ball that you used to introduce the activity, but in that example it is much less clear that something moved while the force was being applied rather than afterwards.

The roller turns as you pull the string, winding up the rubber bands. Now the toy has potential energy. When the string is released, the stored energy is converted to kinetic energy. As the toy moves across the floor, the friction between the rubber bands and the floor converts the kinetic energy to heat. Some energy may also be stored back in the rubber band if the band twists up in the opposite direction. When all of the kinetic energy has been converted to heat or re-stored in the rubber band, the toy stops. None of the toy's original energy has been lost; it has simply changed form. The law of conservation of energy states that in any given situation, energy may change from one form to another, but the total amount of energy remains constant. Energy may change form but it cannot be created or destroyed under ordinary conditions.

The distance the toy moves is proportional to the amount of energy you stored in it originally. The farther out you pull the string, the greater the distance traveled. Do not expect the students to be able to make an exact prediction of how far the toy should travel in step 6 of the worksheet. However, if they select a length less than half the string, they should realize it will go a shorter distance than in step 5. If they select a length greater than half the string, then they should predict a distance in between their previous two measurements.

VARIATION

Have students make papier mâché ladybugs. To make a bowl-shaped mold, blow up a small balloon to about 6 inches in diameter. Wrap the balloon in a piece of wax paper or plastic wrap. Apply papier mâché around the entire surface of the balloon. Cut the papier mâché sphere in half after it has dried completely. Remove the wax paper or plastic wrap. Each half will make one ladybug. Use the Ladybug Assembly Instructions to complete the project, replacing the plastic bowl with the papier mâché form. Decorate as desired.

EXTENSION

Challenge students to design their own version of this energy toy. Toys can be decorated in any fashion, and students can use any type or size of roller. Have contests with various categories: fastest, farthest, biggest, smallest, most comical, etc. Reserve the gym or cafeteria to display and test toys.

ASSESSMENT

Demonstrate for the students another toy that stores energy in a rubber band, such as a wind-up airplane. Ask the students to a) write a paragraph explaining the energy conversions that take place and b) predict what will change if the rubber band is wound more tightly and explain why they think so.

CROSS-CURRICULAR INTEGRATION

Earth science:
- Challenge the students to construct their ladybugs out of all recycled materials. Then have a class discussion about reuse and recycling.
- Real ladybugs are used by many farmers and gardeners to control aphid infestation in their crops. What might be some advantages to using ladybugs for pest control?

Language arts:
- Write a poem or short story about your ladybug. Include a description of the motion of the ladybug.
- Pretend you are a toy manufacturer. Write an advertisement that would attract children to buy your toy. Use something you have learned about motion or energy in your ad (for example, the relationship between work and distance traveled). Name your toy and decide on a selling price for the toy.

- Read aloud or have students read one or more of the following books. Older students could read to a younger reading buddy.
  - *The Grouchy Ladybug,* by Eric Carle, (HarperCollins, ISBN 0064434508)
    A grouchy ladybug looking for a fight challenges everyone she meets regardless of their size or strength.
  - *What about Ladybugs?,* by Celia Godkin, (Sierra Club, ISBN 0871565498)
    A gardener upsets the natural balance in his garden by using poison and learns the value of another method of controlling pests.
  - *Ladybug on the Move,* by Richard Fowler, (Harcourt Brace Jovanovich, ISBN 0152004750)
    Ladybug searches for a new home, but every place she finds is already occupied. Features a separate ladybug who weaves her way through die-cut pages.

Life science:
- Research and write about the function of ladybugs in the ecosystem. What do they eat? Does anything eat them? What habitat do they prefer?

FURTHER READING

Catherall, E. *Exploring Uses of Energy;* Steck-Vaughn: Austin, TX, 1990. (Students)

Challoner, J. *Eyewitness Science: Energy;* Dorling Kindersley: New York, NY, 1993. (Students)

Gartrell, J.E.; Schafer, L.E. *Evidence of Energy: An Introduction to Mechanics;* National Science Teachers Association: Washington, D.C., 1990. (Teachers)

Kirkpatrick, L.D.; Wheeler, G.F. *Physics: A World View,* 2nd edition; Saunders: Philadelphia, PA, 1995. (Teachers)

CONTRIBUTORS

Faye Flavin, Miami University undergraduate student, Hamilton, OH, 1995.

Rebecca Gilbert, Fairbanks Catholic School, Fairbanks, AK, Teaching Science with TOYS, 1996.

Mary Yarger, Sherman Indian High School, Riverside, CA, Teaching Science with TOYS, 1996.

Renita Strange, Springdale Elementary, Princeton, OH, Teaching Science with TOYS, 1996–97.

HANDOUT MASTERS

Masters are provided for the following handouts:
- Ladybug Assembly Instructions
- Data Sheet

Copy as needed for classroom use.

Ladybug, Ladybug, Roll Away

Ladybug Assembly Instructions

1. Place the bowl upside down on a table. Put one fastener through each hole on the sides of bowl. Turn the bowl over and flatten out the metal tabs to hold them in place.

2. Complete the step below that matches the materials you have.

   a. If you are using a **dowel rod**, cut the two small rubber bands so that you have two straight pieces of rubber band. Thread one rubber band through each end of the dowel rod and tie the ends in a knot. Check to make sure both knots are tight. (See Figure 1.)

   Figure 1: For a dowel rod roller, thread one rubber band through each end of the dowel rod and tie the ends in a knot.

   b. If you are using a **thread spool**, cut the long rubber band so that you have one long piece. Slip the rubber band completely through one of the pie-shaped honeycomb sections of the inside of the spool and back through the opposite pie-shaped section. Tie the ends of the rubber band in a knot. (See Figure 2.)

   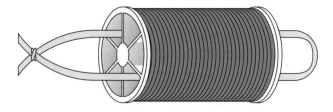

   Figure 2: For a thread spool roller, slip one end of a cut rubber band through one honeycomb section and back through the opposite section.

   c. If you are using a **film canister**, cut the long rubber band so that you have one long piece. Thread both ends of the rubber band through the holes in the bottom of the canister, then through the holes in the lid. Tie a knot in the rubber band. Fill the canister with popcorn kernels and put the lid on it. (See Figure 3.)

   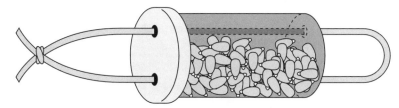

   Figure 3: For a film canister roller, slip the ends of a cut rubber band through the bottom of the canister, then through the holes in the lid.

3. Wrap one or more very wide rubber bands evenly around each end of your roller (dowel rod, spool, or film canister) to provide extra friction. (See Figure 4.)

Figure 4: Wrap rubber bands around each end of your roller to provide extra friction.

4. Tie one end of the string or floss tightly around the center of the roller. Tape the string or floss to the roller to keep it from slipping.

5. Thread the other end of the string or floss through the top hole of the bowl from the inside of the bowl to the outside.

6. Turn the bowl over and tie a plastic ring or button to the free end of the string or floss. (See Figure 5.)

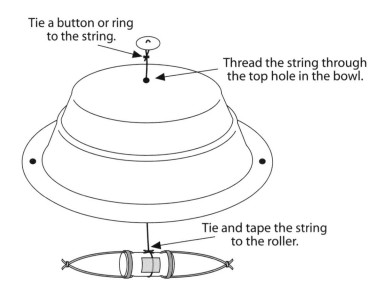

Tie a button or ring to the string.

Thread the string through the top hole in the bowl.

Tie and tape the string to the roller.

Figure 5: Start to assemble the Ladybug toy.

7. Flip the bowl back over, wind the string or floss tightly around the roller until the ring or button is tight against the top of the bowl.

8. Attach each end of the rubber band that was threaded through the roller to the metal paper fasteners by slipping the band under the metal tabs. (See Figure 6.)

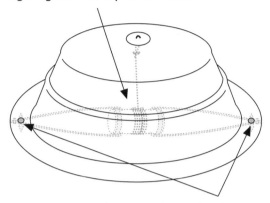

(a) Wind the string around the roller until the button is tight against the top of the bowl.

(b) Fasten the rubber bands by slipping them under the metal paper fasteners.

Figure 6: Finish assembling the Ladybug toy.

9. Set the toy on the floor with the roller touching the floor. Hold the ladybug with one hand, pull the string or floss, and let go of the toy. Try using the toy in different ways and see what it does. Decorate as instructed by your teacher.

Troubleshooting tips: If your ladybug goes in a circle, adjust the traction rubber bands. If too much string does not wind back onto the roller, shorten the string or pre-wind the rubber bands slightly before attaching them to the fasteners. If using the canister model, be sure the canister is full of popcorn kernels.

Name _____   Date _____

Ladybug, Ladybug, Roll Away
Data Sheet

1. Put the toy on the floor. Hold the ladybug with one hand and use your other hand to pull the string, but don't let go yet. Briefly describe what happens when you pull the string.

   _____

   _____

2. Now, let go of the string. In what direction does your ladybug move?

   _____

   _____

3. Turn the ladybug over so that you can see the bottom of the toy. Hold the toy in one hand, pull the string with the other hand, and let go. Describe what happens.

   _____

   _____

4. Now, put the ladybug on the floor again. Place a small piece of tape just in front of the toy. Pull the string out as far as it will comfortably go. Measure the length of the string coming out the top of the toy. Let go. Measure the distance between the piece of tape on the floor and the spot where the toy stopped and record your measurement.

   The string is _____ centimeters long.

   The toy moved _____ centimeters.

5. Repeat step 4, but this time pull the string out half as far as you did the first time. Measure and record the distance the toy travels.

   The string is pulled out _____ centimeters.

   The toy moved _____ centimeters.

Data Sheet, page 2

6. Choose another length of string to try. Before you try it, predict how far you think the toy will go.

If I pull the string out _____ centimeters, then

I predict the toy will travel _____ centimeters.

What information did you use to make your prediction?

_____

_____

_____

Now try it and record your results.

The string is pulled out _____ centimeters.

The toy moved _____ centimeters.

7. What conclusions can you draw about the relationship between how far the string is pulled out and the distance your toy moves?

_____

_____

_____

_____

8. How does your toy acquire the ability to move?

_____

_____

_____

9. What position is the string in when the toy has the greatest amount of potential energy?

_____

_____

10. When do you think the toy has the most kinetic energy?

_____

_____

11. Starting with when you pull the string, describe all the energy transformations that take place in the operation of this toy.

_____

_____

_____

_____

_____

_____

_____

_____

_____

_____

Rubber Band Airplane

...In this activity, students investigate energy storage in a wingless airplane.

✔ **Time Required**

Setup	20	minutes
Procedure	50	minutes
Cleanup	5	minutes

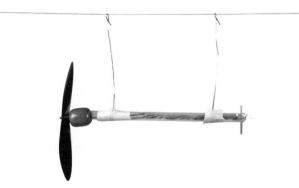

A rubber band airplane

✔ **Key Science Topics**

- elastic potential energy
- force
- friction
- kinetic energy
- Newton's third law

✔ **Student Background**

Students should have completed some activities on kinetic and potential energy. Students should have previously studied Newton's third law of motion, although that part of the discussion could be omitted if they have not.

✔ **National Science Education Standards**

Science as Inquiry Standards:

- Abilities Necessary to Do Scientific Inquiry

  Students form a hypothesis and design and execute an experiment to test their hypothesis. Students graph and interpret their data and present evidence for their conclusions.

Physical Science Standards:

- Motions and Forces

  The unbalanced force of the air on the propeller causes the toy to begin moving. After the propeller stops turning, the force of friction on the support wires causes the toy to stop.

- Transfer of Energy

  Energy can be stored in rubber bands and transferred to the toy.

MATERIALS

For the Procedure
Per group of 3–4 students
- small paper clip
- 25-cm piece of string or thread
- plastic propeller about 10 cm long

➤ *These propellers are available from most hobby stores or can be ordered from Kelvin, 10 Hub Drive, Melville, NY 11747; 800/535-8469.*

- 2 pieces of thin, smooth wire about 20 cm long each (such as 24-gauge galvanized steel wire)
- 2 medium-sized rubber bands of the same size (Number 18 works well.)
- small plastic craft bead

➤ *Almost any shape will work, but the bead must be larger in diameter than the straw, and the hole in the middle must be big enough to poke the straightened paper clip through.*

- straw

➤ *Clear ones are ideal because students can see the rubber band inside.*

- tape
- toothpick
- 5–6 m monofilament fishing line
- meterstick or other device for measuring distance
- scissors
- (optional) pliers

➤ *Younger students may need these for bending the paper clip, but older ones may not.*

For the Extension
Per class
- Delta Dart kit

➤ *Delta Dart kits are available from Midwest Products Co., Inc.; 400 S. Indiana Street; P.O. Box 564; Hobart, IN 46342; 219/942-1134.*

GETTING READY

Plan where the tracks can be set up and what the lines can be tied to. If necessary, rearrange the furniture to provide space for each group to set up its track. Set up one track by stretching 5–6 m fishing line between two tables or chairs; be sure the track is level and pulled tight.

Build one airplane for demonstration according to the Airplane Assembly Instructions sheet (provided) and determine the lowest number of winds that will cause the airplane to move. In most cases, it will be around 20–30, but the number will depend on the strength of your rubber bands.

INTRODUCING THE ACTIVITY

Show the students a completed airplane. Wind it up, place it on the track, and ask them what they think will happen when it is released. Release the airplane. Ask the students to explain what happened in terms of energy and what force moved the airplane forward. (If they have no previous experience with propellers, they may not be able to do this. In that case, instruct them to pay close attention to the shape of the propeller and how it interacts with the air as they do the experiment. During the closing discussion, return to this issue of what force moves the plane. See the Explanation for a brief discussion of how propellers work.)

PROCEDURE

1. Divide students into groups of three or four and have each group build an airplane following the instructions on the Airplane Assembly Instructions sheet.

   You could save class time by having a few students build the airplanes in advance.

2. Explain to the students that you want them to investigate whether different amounts of energy can be stored in the airplane. Ask students to form a hypothesis about the outcome of the investigation based on other experiments they have done, such as the Explorer Gun® experiment.

3. Ask students to design an experiment to test their hypothesis.

   One possible method is to start with the lowest number of turns that will make the airplane move and measure how far along the track it moves. Be sure to determine the initial and final positions using the same point on the airplane. Then repeat the measurement, increasing the number of turns by 10 each time.

4. Have students carry out the experiments they designed. They may graph their data if you wish.

5. Ask students what conclusions they can draw from their data. Be sure to have them clearly state what evidence supports that conclusion.

6. Ask students whether their plane stores energy like an Explorer Gun disk in the activity "Exploring Energy with an Explorer Gun" or a See Thru Racer in the activity "How Much Energy?" (This question, which addresses linear and non-linear relationships, may be easier to answer if students have graphed their data.)

EXPLANATION

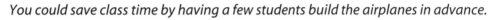

The following explanation is intended for the teacher's information. Modify the explanation for students as required.

The force that makes the airplane move can be explained in terms of Newton's third law: When object A exerts a force on object B, then object B exerts a force of equal size and in the opposite direction back on object A. The slanted propeller blade exerts a force on the air that is both downward and backward. The air then must be exerting a force upward and forward on the propeller. This forward force moves the airplane. If you hold your hand behind a propeller while it is turning, you will feel a wind. This is the air that is being pushed backwards by the propeller.

The rubber band provides the energy to make the airplane move. When you wind up the rubber band, you are doing work. You transfer chemical energy from your body to the rubber band. The energy is stored as elastic potential energy until the propeller is released and the rubber band unwinds. When you release the propeller, the rubber band tries to unwind, exerting a force on the propeller that causes it to turn. Eventually all of the elastic potential energy is transformed into the kinetic energy of the airplane.

The force of friction between the wire hangers and the fishing line eventually converts all the airplane's kinetic energy into thermal energy, and the airplane stops. The distance the airplane travels before it stops is a measure of the maximum amount of kinetic energy the airplane had.

The students will find that more energy can be stored in the rubber band by winding it more. This concept is shown by the fact that the airplane travels farther with increasing numbers of winds. They will probably find that the relationship between number of winds and distance traveled is not linear. That is, each increase of 10 winds does not produce the same amount of increase in the distance traveled. In this respect, the rubber band airplane is more like the Explorer Gun than the See Thru Racer. If students graph their data, it will not fall in a straight line. See Figure 1 for a sample graph.

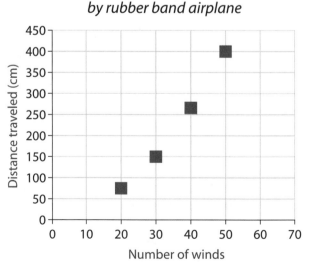

Figure 1: Effect of winding on distance traveled by rubber band airplane

EXTENSIONS

Use this activity as a springboard into the study of aerodynamics. Build a simple balsa wood plane driven by a rubber band, such as the Delta Dart. A lesson plan that uses this toy can be found in the *Teaching Physics with TOYS* activity "Delta Dart."

ASSESSMENT

Have the students complete the Assessment Sheet (provided).

CROSS-CURRICULAR INTEGRATION

Language arts:

- Have students read the following fiction book:
  - *Dragonwings,* by Laurence Yep (HarperTrophy, ISBN 0064400859)
    In the early 20th century a young Chinese boy joins his father in San Francisco and helps him realize his dream of making a flying machine.

- Have students read one or more of the following nonfiction books:
  - *Flight: The Journey of Charles Lindbergh,* by Robert Burleigh (Philomel, ISBN 0399222723)
    Describes how Charles Lindbergh achieved the remarkable feat of flying nonstop and solo from New York to Paris in 1927.
  - *The Story of Flight: Early Flying Machines, Balloons, Blimps, Gliders, Warplanes, and Jets* (Scholastic, ISBN 0590476432)
    Chronicles the history of flight, examining how the trials and triumphs of early aviators led to the high-speed air travel of today.
  - *Airplanes,* by Steve Parker (Copper Beech, ISBN 156294311X)
    An imaginative look at airplanes through such questions as "What if airplanes didn't have air?" and "What if airplanes could go into space?"
  - *Eureka! It's an Airplane!,* by Jeanne Bendick (Millbrook Press, ISBN 1562940589)
    Describes the development of the airplane and some of the inventions that have made it a common means of transportation.
  - *The Wright Brothers: How They Invented the Airplane,* by Russell Freedman (Scholastic, ISBN 0590464248)
    Follows the lives of the Wright brothers and describes how they developed the first airplane.
  - *The Wright Brothers at Kitty Hawk,* by Donald Sobol (Scholastic, ISBN 0590464248)
    Tells the story of the Wright Brothers' efforts to build a successful flying machine.

- For a creative writing activity, have students write a letter to the Wright Brothers discussing the brothers' new invention, the airplane.

- Have students research the Wright Brothers and create a journal that one of the Wright Brothers could have written telling how they invented the airplane or tested the airplane.

Math:
- Have students graph the data they collected.

FURTHER READING

Millspaugh, B.; Taylor, B. *Let's Build Airplanes and Rockets;* Learning Triangle Press: New York, NY, 1996. (Teachers and Students)

CONTRIBUTORS

Glen Gillen, Miami University graduate student, Oxford, OH, 1996.

Terry Skudlarek, Mound Elementary School, Miamisburg, OH; Teaching Science with TOYS, 1996–97.

HANDOUT MASTERS

Masters for the following handouts are provided:
- Airplane Assembly Instructions
- Assessment Sheet

Copy as needed for classroom use.

Rubber Band Airplane

Airplane Assembly Instructions

1. Open the paper clip and straighten one end, leaving the other end in a hook shape as shown in Figure 1. Use the pliers as necessary.

Figure 1: Unbend the paper clip into this shape.

2. Slide the propeller and craft bead onto the paper clip from the straight end. (See Figure 2.)

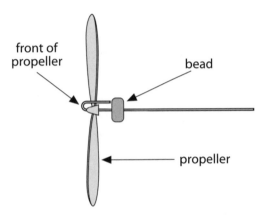

front of propeller

bead

propeller

Figure 2: Slide the propeller and craft bead onto the paper clip.

3. Bend the other end of the paper clip back to its original position. The bend needs to be tight enough that it will fit inside the straw without rubbing the straw as it spins. (See Figure 3.)

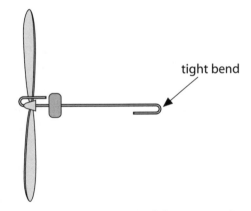

tight bend

Figure 3: Bend the end of the paper clip.

4. Tie the two rubber bands together with the piece of string or thread. Hook the rubber bands onto the paper clip. (See Figure 4.)

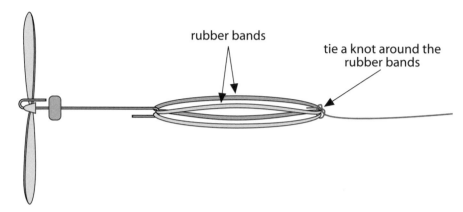

Figure 4: Tie the rubber bands together and place them on the paper clip.

5. Lay the assembly on a desk or table and straighten all the pieces. Lay the straw alongside the assembly with the top of the straw just below the bead. Mark the straw ½ to ¾ inch longer than the end of the rubber bands with a marker and cut the straw at that mark. (See Figure 5.)

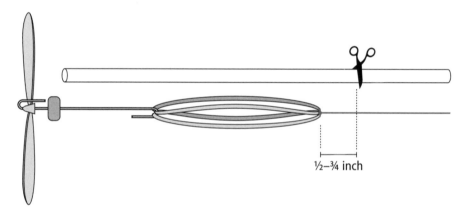

Figure 5: Mark and cut the straw about ½–¾ inch from the end of the rubber bands.

6. Cut two small notches about ⅛ inch long opposite each other on one end of the straw. (See Figure 6.)

Figure 6: Cut notches opposite each other on one end of the straw.

Reproducible page from *Exploring Energy with TOYS* published by Terrific Science Press™

7. Thread the string end through the straw, starting at the end of the straw without the notches. (See Figure 7.)

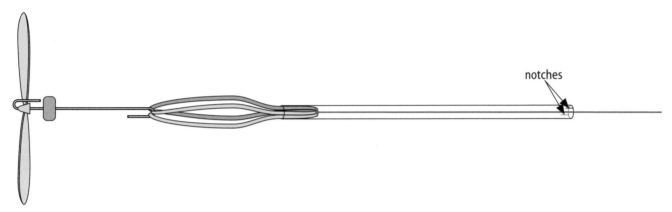

Figure 7: Thread the string and rubber band through the straw.

8. Pull the rubber bands through the straw so the craft bead sits on one end and the rubber bands stick out of the other end. (See Figure 8.)

Figure 8: Pull the rubber bands through the straw.

9. Put the toothpick through the two rubber bands where they stick through the straw and rest the toothpick inside the notches at the end of the straw. (See Figure 9.)

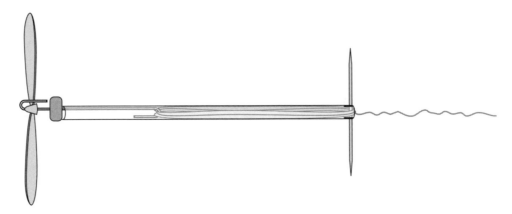

*Figure 9: Feed the toothpick through the rubber band
and rest the toothpick in the notches.*

10. Make two hangers by bending the pieces of wire around the straw. Be careful not to bend the wires so tightly that they pinch the straw. Tape the wires in place on the straw. Bend the free end of each hanger into a hook. The hangers should be the same height, and the wire should be as straight as possible between the hooks and the straw. (See Figure 10.)

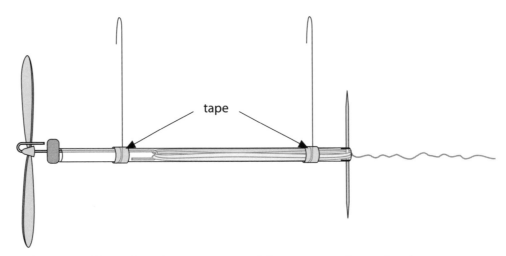

Figure 10: Wrap the wire once around the straw and tape in place.

11. Make a track for the airplane by tying a 5- to 6-m length of fishing line to two solid objects of equal height, such as chair backs. Make sure this track is level and pulled tight so it doesn't sag.

12. Hook the airplane on the fishing line "track." The airplane should be level when hung on the track. (See Figure 11.) If it is not level, bend one of the wires to change the height of the hook. When the airplane is level, twist the hooks closed.

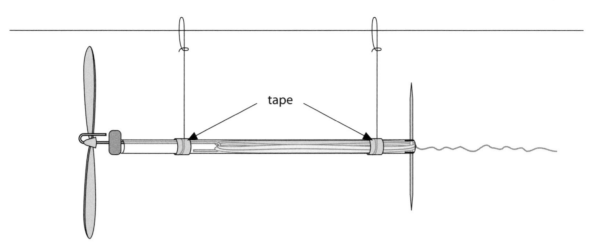

tape

Figure 11: Wrap the wire once around the straw and tape in place.

13. Rotate the propeller, holding the toothpick in place if necessary. Release the propeller. If the airplane does not move forward, try winding the propeller in the opposite direction. If the airplane still does not move forward, try winding the propeller several more turns. If the propeller does not turn at all after the rubber band is wound, check the spot where the straw touches the craft bead. The hole in the bead should be small enough that it does not catch the edge of the straw. If the hole is too big, try another bead.

Name _____   Date _____

Rubber Band Airplane
Assessment Sheet

1. Draw a picture of your rubber band airplane in the space below. Label its parts.

2. What happened to the rubber band when you wound the propeller?

   _____

   _____

   _____

3. When you let go of the propeller, what did the rubber band do?

   _____

   _____

   _____

4. What kind of energy is created when the rubber band is twisted? Write a statement about this kind of energy.

   _____

   _____

   _____

   _____

   _____

   _____

Reproducible page from *Exploring Energy with* **TOYS** published by Terrific Science Press™

5. Describe the energy the airplane had after the propeller was let go. What is this type of energy called?

_____

_____

_____

6. What does the rubber band do besides store energy?

_____

_____

_____

7. How does the motion of the propeller cause the airplane to move?

_____

_____

_____

_____

8. Apply what you learned in this activity to how a real propeller-driven airplane works. Discuss similarities and differences.

_____

_____

_____

_____

_____

_____

_____

Slingshot Physics

...Students discover the relationship between the distance the slingshot is stretched and the distance the ball travels.

✔ Time Required

Setup   10–30   minutes*
Performance   100   minutes (two 50-minute periods)
Cleanup   5   minutes

* Depending on whether the students or the teacher prepares the straws

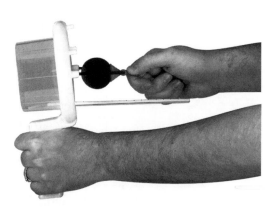

Sling Thing

✔ Key Science Topics

- elastic potential energy
- kinetic energy
- work

✔ Student Background

Students should have been introduced to the concepts of work, potential energy, and kinetic energy. If the students have not encountered these ideas in earlier grades, use the activity "What Makes It Go?" to introduce them.

✔ National Science Education Standards

Science as Inquiry Standards:

- Abilities Necessary to Do Scientific Inquiry

  Students identify variables that must be controlled during the experiment.

  Students use graphs to interpret their data and make predictions based on experimental evidence.

Physical Science Standards:

- Motions and Forces

  The force of the cord on the ball causes the ball to increase its speed.

  The downward force of gravity on the ball causes it to change its direction of motion.

- Transfer of Energy

  Energy is associated with the motion of the ball and the stretching of an elastic cord. As the cord is stretched greater distances, more energy is stored.

✔ Additional Process Skill

- collecting data   Students will measure the distance the ball travels as a function of the distance the launcher is stretched.

MATERIALS

For Getting Ready
Per class
- ruler
- tape
- permanent fine-tipped marker

Per group

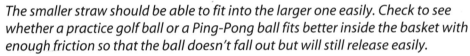

While the directions are written for either the Sling Thing or the Water Bomb Sling Shot, the activity could be modified to use any slingshot-type toy.

- 1 of the following:
  - Sling Thing™, soft foam ball "ammunition," and 1 plastic drinking straw
  - Water Bomb Sling Shot, "wiffle"-type practice golf ball or Ping-Pong™ ball, and 2 drinking straws of different diameters

The smaller straw should be able to fit into the larger one easily. Check to see whether a practice golf ball or a Ping-Pong ball fits better inside the basket with enough friction so that the ball doesn't fall out but will still release easily.

For the Procedure
Per group
- Sling Thing or Water Bomb Sling Shot with straw(s) attached (See Getting Ready.)
- measuring device, such as a meterstick, measuring tape, or trundle wheel
- graph paper
- (optional) measuring stick (if student desks or chairs are not available during experimentation)

For the Extensions
❷ All materials listed for the Procedure plus the following:
Per group
- different sizes and masses of balls

SAFETY AND DISPOSAL

Remind the students not to launch balls at people or breakable objects. No special disposal procedures are required.

GETTING READY

These steps could be done by the students rather than the teacher. However, the more carefully they are done, the more reliable the data will be, so you should judge whether your students have both the skill and patience to do them correctly. Leave one slingshot unmodified for initial demonstration.

If using the Sling Thing, follow these steps:

1. Mark a straw in intervals of 1 cm. When this straw is attached to the Sling Thing in step 2, it will help the Launcher to determine how far back he or she has pulled the elastic cord, and thus, how much work was done on the toy.

2. Draw a mark on the center of the prong on the Sling Thing with a permanent marker. (See Figure 1.) Students will slide the ball back to this mark, and they will use the mark to measure pull-back distance.

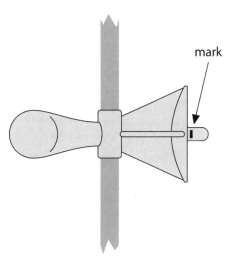

Figure 1: Draw a mark at the center of the prong.

3. Line up one of the marks on the straw directly under the elastic cord. (See Figure 2.) Use tape to attach the marked straw to the bottom of the safety shield under the elastic cord. Write a "0" on the straw at this mark.

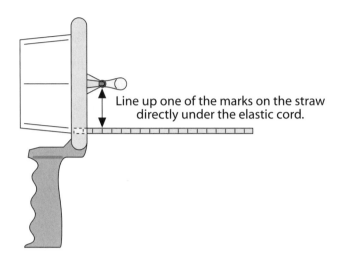

Figure 2: Line up the straw and elastic cord and tape the straw to the Sling Thing.

If using the Water Bomb Sling Shot, follow these steps:

1. The Water Bomb Sling Shot is designed to throw water balloons, but these are extremely messy and do not provide consistent data, so use practice golf balls or Ping-Pong balls instead. Tape the smaller drinking straw across the crossbar of the Sling Shot so that it protrudes back toward the shooter horizontally as shown in Figure 3.

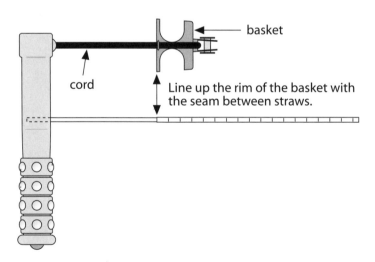

Figure 3: Set up the Water Bomb Sling Shot as shown.

2. On the larger straw, mark equal intervals of 1 cm beginning at one end. Slide this end of the marked straw over the end of the taped straw, overlapping them about 5–6 cm.

3. Pull the basket straight until the elastic is taut but not stretched. Adjust the seam between the straws so that it is even with the rim of the basket. Tape the straws together and draw a line marked 0 cm on the seam as shown in Figure 3. The 0-cm line indicates zero energy.

Since the rim of the basket is used to determine the zero point, it should also be used to determine the pull-back distance. Other points on the basket could be used, but only one should be used for a given toy.

Testing the Slingshot

Determine the minimum and maximum distance for which it is practical to stretch the cord of the toy being used. For both the Sling Thing and the Water Bomb Sling Shot, the minimum distance that will launch the ball is about 3 cm. The maximum pull-back distances of the Sling Thing and the Water Bomb Sling Shot will vary from toy to toy but will probably not be more than 12 cm and 18 cm, respectively. Decide which pull-back intervals you want students to use to collect data. For example, you can have students start at 3 cm and collect data every 2 or 3 cm up until your maximum pull-back distance. Also decide how many trials students will do for each interval.

INTRODUCING THE ACTIVITY

Announce to the students that today paper wads are legal. Set up a few trash cans as targets and give a few students some paper to crumple into wads. Ask students to propose ways to get the paper wads into the cans. As each method is proposed, ask a student with a paper wad to demonstrate that method. For instance, it could be just carried over and dropped in, thrown, kicked, or batted with a book. Once someone proposes shooting it with a rubber band, have a student demonstrate this method. Ask how shooting the paper with a rubber band is different from the other methods. *Energy is stored before being transferred to the paper wad.*

PROCEDURE

Part A: Conducting the Investigation

1. Demonstrate the operation of an unmodified slingshot toy. Demonstrate that the elastic cord can be pulled back to different intervals. Ask the students to ponder the effects that the elastic cord has on the operation of the toy. "Can we design a test to see whether it matters how far back we pull the cord? Will that affect the distance that the ball travels?" Discuss possible tests. Point out with the students' help what things must be kept constant in order to make the test "fair" every time. List these on the board. Possible variables include the following:

   • angle at which the slingshot is held

- height from which the ball is launched

- mass of the ball

- conditions in the environment, such as wind

- size of the hole in the ball (for Sling Thing)

2. Take the students to a location with sufficient room for all groups to carry out the experiment at the same time.

3. Go through the plan that you have created for the experiment. Students will be measuring the distance between the launch and landing points for various amounts of stretch of the cord. The Launchers (see step 4) should be careful to use the same technique each time. It is very important that the cord be pulled straight back horizontally, not at an angle, every time. All groups should launch from the same height. A convenient way to do this is to rest the handle on a tabletop, chair back, or meterstick.

4. Divide the class into groups of four.

Each group will need a Launcher, Recorder, Spotter, and Measurer to do the following jobs:
- *Launcher—to launch the ball;*
- *Recorder—to record all measurements on the Data Sheet (provided), to help the Launcher achieve the correct pull-back distance, and to assist the Launcher in making sure that the launch is horizontal;*
- *Spotter—to mark where the ball hits first, to retrieve the ball, and to assist in measuring the distance; and*
- *Measurer—to measure the distance the ball travels and relay the measurements to the Recorder.*

5. Have the Launcher launch the ball for the first interval being tested. He or she should carefully pull the cord back to the desired interval marked on the straw, with the assistance of the Recorder.

You may want to have the Launcher do a few practice launches before beginning the experiment.

6. Have the Spotter mark where the ball hits and retrieve the ball.

7. Have the Measurer, with the assistance of the Spotter, measure the distance traveled by the ball. The distance traveled should be measured on the floor starting directly below the zero point on the straw (the position of the ball when the cord is not stretched). Have the Recorder record the distance on the Data Sheet.

8. Have each group repeat steps 5–7 for as many trials as desired for each interval being tested. You can shorten the amount of time spent on testing by having each group complete only one trial at each interval. In this case, you should omit the group graphs described below and just do a class graph.

Part B: Compiling Data

1. The following class period, have each group calculate averages for each of the intervals that were tested and create a graph of their data. The bar or line graph will show the distance that the ball traveled for each interval that the cord was pulled back. For sample data, see the Sample Data sheet (provided). This sheet is for your reference only; students should make their own tables and graphs.

2. Post the graphs and have one student from each group list the group's data in a class chart. As a class, check class data for consistencies and inconsistencies.

3. Calculate a class average for each interval and create a class graph of this data.

CLASS DISCUSSION

1. Ask students to explain how the toy works. (They should be able to explain that the Launcher does work on the toy as he or she pulls the cord back. This energy is stored in the elastic cord as elastic potential energy and is converted to kinetic energy when the cord is released. The ball is launched when the cord reaches its physical limit of motion. The ball then continues to travel horizontally at the same speed while the force of gravity pulls it downward.)

2. Ask the students to look at the graphs and decide whether the distance the cord is stretched affects the distance traveled. Ask them to relate this to the energy ideas just discussed. *The farther the cord is stretched, the more energy is stored and later converted to the kinetic energy of the launched ball.* Ask the Launchers how they know more work was done to stretch the cord greater distances. *They had to use more force to pull the cord back.* Discuss how the graph could be used to predict the distance traveled for pull-back distances they did not test. Discuss how this process is different for linear and non-linear graphs. (See the Explanation.) Have students complete the "Drawing Conclusions from the Data" section of the Data Sheet.

3. Look at the class data and discuss why more data are better in a test like this. Discuss how factors students were unable to control might have influenced individual trials.

EXPLANATION

The following explanation is intended for the teacher's information. Modify the explanation for students as required.

When you pull back the cord, you do a certain amount of work. The farther back you pull the cord, the more force you are exerting, and the more work you do. Energy from your body is stored in the elastic cord as elastic potential energy. When you release the cord, the cord begins to move in the opposite direction from the one in which you pulled it; thus, elastic potential energy changes to kinetic energy.

Physical limits restrict how far the cord can move in either direction, but the ball has no limits on how far it can move. During launching, when the cord (with holder and ball) has sprung forward as far as it can go, it snaps back toward the person holding the toy, but the ball continues moving forward, away from the shooter. The ball is able to fly off (or out of) the holder because the inertia of the ball overcomes the friction that initially kept it on (or in) the holder.

The initial velocity with which the ball leaves the holder depends on the amount of kinetic energy it has received from the cord. The velocity of the ball as it leaves the toy determines the distance the ball travels before hitting the ground, assuming the time the ball stays in the air is constant. This time will depend on how far from the ground the ball is launched; thus, the launch height must remain constant for each trial.

In this activity, ideally the only variable being tested is the interval to which the cord was pulled back. However, hidden variables may come into play that are difficult to control, such as slight differences in the technique of the Launchers, slightly different strengths of the elastic cord from one toy to another, wind, or differences in the size or shape of the balls. The most important of these hidden variables is the launch itself. If the ball is launched at an upward angle (up to 45°), the distance will be increased.

An easy way to see the relationship between cord pull-back and distance traveled is to graph the data. Besides making trends easier to see, graphs also help us to predict what will happen in other experiments. In a linear graph, equal changes on the horizontal axis correspond to equal changes on the vertical axis. For instance, in this experiment, data would be linear if all 1-cm increases in how far the cord was pulled back resulted in equal increases in distance traveled. If the largest pull-back distance was 12 cm, and students wanted to predict how far the ball would travel with a pull-back distance of 13 cm, they would simply add on the standard increase. However, the data may be non-linear—equal changes on the horizontal axis may not correspond to equal changes to the vertical axis. To use non-linear graphs for predictions, students would draw a smooth curve through the data points so they could see where additional points within the same range would fall, based on the trend.

EXTENSIONS

1. Have the students compare results for the same amount of stretch but different launch heights. Does height affect the distance traveled?

2. Have students test different sizes and masses of balls with the Water Bomb Sling Shot. (This will not work well with the Sling Thing.) Does a more or less massive ball travel differently with the same initial energy? Does the size of the ball change the distance traveled at each interval?

ASSESSMENT

Options:

- Collect and evaluate the "Drawing Conclusions" section of the Data Sheet.

- Have the students explain in writing how shooting paper wads with a rubber band is similar to using a Sling Thing or Water Bomb Sling Shot.

- Have students draw designs of slingshots made with unusual materials. Have them explain how they used what they learned in the activity in making their choices of materials.

- Have the students list and explain two reasons why graphs are very useful in scientific tests.

CROSS-CURRICULAR INTEGRATION

Language arts:
- If students regularly keep science journals, ask them to write about what they learned about slingshots in this activity. Also, ask them to list safety procedures one should follow when using slingshots.

Math:
- Use this activity as a math activity focusing on measuring, averaging, and graphing, with science as the secondary focus.

Physical education:
- Have students use their slingshots to shoot at a target. Periodically change the location of the target. Discuss the factors that are important for accuracy.

Social studies:

- Research the early history of the slingshot. How old are the oldest slingshots that have been found? How did different cultures use them? Are they still important in some cultures?

FURTHER READING

Kirkpatrick, L.D.; Wheeler, G.F. *Physics: A World View,* 2nd ed.; Saunders: Philadelphia, PA, 1995. (Teachers)

CONTRIBUTORS

Lee Ann Ellsworth, Northwestern Middle School, Springfield, OH; Teaching Science with TOYS, 1991–92, Research Teacher, 1995.

Sue Hoane, Westwood Middle School, Grand Rapids, MI; Teaching Science with TOYS, 1996.

Teresa Mangen, New Lebanon Middle School, New Lebanon, OH; Teaching Science with TOYS, 1995.

Darlene Tucker, Thomas Jefferson Middle School, Menomonee Falls, WI; Teaching Science with TOYS, 1996.

HANDOUT MASTERS

Masters for the following handouts are provided:

- Sample Data
- Data Sheet

Copy as needed for classroom use.

Slingshot Physics

Sample Data

Sling Thing Data				
Pull-Back Distance (cm)	Trial 1 (cm)	Trial 2 (cm)	Trial 3 (cm)	Average (cm)
3	63	137	76	92
4.5	185	110	85	127
6	178	212	157	182
7.5	312	321	242	291
9	303	263	211	259
10.5	336	297	416	350
12	498	359	281	379

Water Bomb Sling Shot Data				
Pull-Back Distance (cm)	Trial 1 (cm)	Trial 2 (cm)	Trial 3 (cm)	Average (cm)
3	127	127	161	138
6	265	311	230	268
9	437	472	322	410
12	529	644	460	544
15	575	690	541	602
18	667	805	621	698

Sample Data Graphs

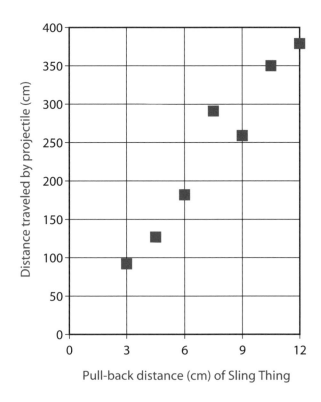

Pull-back distance (cm) of Sling Thing

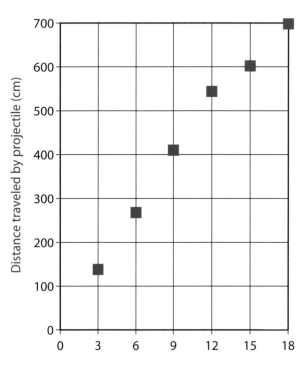

Pull-back distance (cm) of Water Bomb Sling Shot

Slingshot Physics
Data Sheet

COLLECTING DATA

Make all measurements in centimeters.

Pull-Back Distance (cm)	Distance Traveled by Ball (cm)			Average (cm)
	Trial 1	Trial 2	Trial 3	

GRAPHING THE DATA

Make a line graph to display your data. Plot the average distance traveled for each pull-back distance. Remember that all graphs should be neat and contain the following:

- An appropriate title

- Labels on both the x and y axes

- Scales on both the x and y axes

DRAWING CONCLUSIONS FROM THE DATA

1. As the pull on the slingshot increases, the distance the ball travels _____ (increases, decreases, or stays the same). This is a _____ (direct or inverse) relationship.

2. Choose a pull-back distance in between two measurements you made, such as 8.5 cm. Predict how far you think the ball would travel for this distance.

   _____

3. Explain your prediction.

   _____

   _____

   _____

   _____

4. If you could pull the slingshot back to 1 cm farther than the longest pull-back distance you tested, how far do you think the ball would travel?

   _____

5. Explain how you made the prediction.

   _____

   _____

   _____

   _____

   _____

   _____

   _____

The Catapult Gun

...Students use an Insect Gun and projectiles of various masses to investigate the relationship between mass and kinetic energy.

✔ Time Required

Setup   5 minutes
Performance   60 minutes*
Cleanup   5 minutes

*Divide into two sessions for younger students.

✔ Key Science Topics

- energy conversions
- kinetic energy
- potential energy

Insect gun and projectiles

✔ Student Background

Students should be familiar with the concepts of potential and kinetic energy and friction.

✔ National Science Education Standards

Science as Inquiry Standards:

- Abilities Necessary to Do Scientific Inquiry

  Students investigate how the mass of a thrown object affects how far it travels.

  Students interpret their data and explain their observations in terms of kinetic energy.

  Students think critically about anomalous data that does not fit the general pattern.

  Students predict the distance another projectile will travel based on the data from their experiment, then test their prediction.

Physical Science Standards

- Transfer of Energy

  Energy can be transferred from one object to another.

✔ Additional Process Skill

- measuring   Students measure the distance traveled by objects shot from the catapult gun.

MATERIALS

For the Procedure
Per group
- Insect Gun

 *This toy is available from Archie McPhee and Company, Seattle, WA; 425/349-3009; fax 425/349-5188; http://www.mcphee.com.*

- plastic insect (Several come with the Insect Gun.)
- 1 of the following sets of projectiles:
  - 1½-inch bolt with 8 nuts
  - clay balls of different sizes
  - set of different-sized fishing sinkers
- piece of Styrofoam™ weighing less than 0.5 g

A small slice of a Styrofoam ball or Styrofoam packing peanut works well.

- small bean such as a navy or pinto bean
- miniature marshmallow

The previous materials are ones that we have tested and found to work well. Other objects may work well, as long as they have obvious mass differences that are spread over a fairly wide range (less than 0.5 g to more than 20 g). Also, at least one object should provide discrepant results, such as Styrofoam. All the objects should be small enough to fit in the catapult tray without touching the sides.

- meterstick or other distance measuring device
- masking tape and pen

Per class
- balance

If your classroom does not have a balance appropriate for measuring small masses, you could ask a high school science teacher to measure the masses of the projectiles for you ahead of time.

For the Extension
All materials listed for the Procedure, plus the following:
Per group
- protractor or file card marked with angles

SAFETY AND DISPOSAL

Take care not to aim the catapult gun at any individual during firing. While it may seem that firing metal objects is dangerous, actually these heavier objects move slowly and do not go very far. Injury is extremely unlikely unless someone is very close to the gun when it is fired.

GETTING READY

Test-fire the objects you have collected. You should have four or five objects for which more mass correlates with shorter distance. This will not be the case for very light objects, such as the piece of Styrofoam. Air resistance (friction with the air) affects objects with low mass much more than objects with high mass. Thus, even though very light objects may be given a large velocity by the gun, air resistance quickly stops them, so they do not go very far.

INTRODUCING THE ACTIVITY

Ask students if they have ever fired a toy gun (such as a Ping-Pong™ ball gun, plastic dart gun, etc.). If they have, ask whether they know what was inside the gun that made it work. (These guns all contain springs of some type.) This can lead the class into a discussion of ways to store potential energy as well as how energy is converted from one form into another.

Next, ask students what would happen if they used a Ping-Pong ball gun, but instead of using a Ping-Pong ball as a projectile they used a Nerf® ball. Accept all answers at this point. Tell the students that you will come back to this question after the activity.

PROCEDURE

Part A: Investigating the Firing Mechanism

Ask students to investigate the firing mechanism of the catapult gun and to discuss in their groups how energy is stored in the catapult gun as well as what energy conversions take place as the catapult gun is fired. After the groups have had time to complete the investigation and discussion, ask one or two groups to report their findings to the class.

Part B: Experimenting with the Catapult Gun

Give students directions for completing the following experiment, in which the mass of the projectile is varied.

1. Review with students the variables that they must control (such as gun height and levelness of the barrel of the gun) in order to make the results of this investigation consistent.

2. Have students use a balance to find the mass of each of the objects and record the masses.

 When measuring small objects, the results will be more accurate if students find the mass of ten objects and then divide that number by ten.

3. Have students examine the available projectiles and predict which will land farthest from the catapult gun when fired.

4. Have students pick an initial firing position on the edge of their desks and mark this point on the floor with a piece of masking tape. They will need about 5 m of shooting range.

5. Have one student in a group hold the catapult gun against the desk in a horizontal position (parallel to the floor) and have a second student view the gun from the side to make sure the gun barrel stays horizontal. Have the first student load and fire one of their collection of objects from the catapult gun. Have a third student mark the landing spot of the object on the floor with a small piece of masking tape. Fire each object from the catapult gun several more times and mark each landing place.

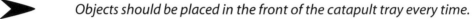 *Objects should be placed in the front of the catapult tray every time.*

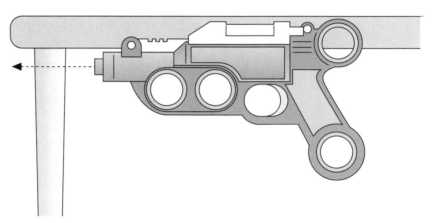

Figure 1: Make sure the gun barrel stays horizontal in relation to the tabletop or desk top.

6. Have students use a meterstick to measure and record the distance in meters from the launch point to each piece of tape. Average the distances the object traveled for all trials.

 Alternatively, instead of averaging distances, you may have students visually estimate the center of the cluster of tape markers and measure the distance from the launch point to this estimated average. This method of averaging may save time.

7. Have students repeat this process with the other objects, using the same starting point. In the case of the nuts and bolt, be sure to shoot the bolt by itself, a nut by itself, and the bolt with varying numbers of nuts screwed onto it.

8. The students should compare their results to their predictions.

9. Ask students to graph the mass of each object versus the average distance each object traveled.

Part C: Analyzing and Interpreting Data

1. When all groups have completed the experiment, ask some of the groups to put their graphs on the board. Ask the other groups if their graphs agree with these in general shape even though their numbers may be different. (For a sample of a typical graph, see Figure 2.) Ask students whether their predictions were correct. Discuss what factors they used to make their predictions. Ask what conclusions can be drawn from the data. Students should note that in most cases the lower the mass of the object, the farther the object travels. Help students connect this observation to the fact that the amount of energy stored in the catapult gun and then converted to kinetic energy is the same each time. Kinetic energy depends on the product of mass and speed, so objects with large mass will have small speed and vice versa. Discuss why the very low-mass object is an anomaly in the data. Discuss the importance of thinking critically about scientific evidence, particularly in accounting for anomalous data. (See the Explanation for further discussion.)

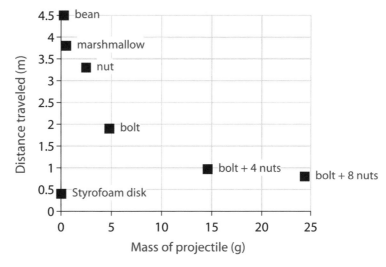

Figure 2: Effect of projectile mass on distance traveled.

2. To conclude the activity, ask students to predict how far one of the insects that came with the catapult gun will travel based on the data they have taken. If they suggest doing so, allow them to determine the mass of the insect before making their predictions. Have them launch the chosen insect and measure the distance. Discuss how close their prediction was to the measurement and how they made their prediction. Also, return to the question about using a Nerf ball in a Ping-Pong ball gun, with which you began the lesson.

VARIATION

This activity could also be done using a homemade catapult. Designs for these can be found in many science activity books.

EXTENSION

Allow students to investigate how the angle at which the catapult gun is held affects how far the projectile travels. Students can design their own experiments, or you can give them the instructions below. If students are not proficient in the use of protractors, you may wish to draw lines at 0°, 15°, 30°, 45°, and 60° on a file card. The angles can be marked as positions 1, 2, 3, 4, and 5. Students can hold the top surface of their catapult gun next to the card to obtain the proper alignment for their launching angles. The card can be held against a tabletop to ensure that the card itself is level and that the height of the catapult gun is always the same. If desired, use the following procedure:

a. Pick one of the projectiles to launch at five different angles.

b. Have one student in each group hold the catapult gun horizontally (0° or Position 1 on their card) and fire the chosen object. A second student can mark, measure, and record the landing position on the floor.

c. Have the launchers line up the top of the catapult gun with the 15° mark on a protractor (or Position 2 on their card) and launch the same projectile. The landing position should again be marked, measured, and recorded.

d. Have students repeat the launch for the other positions and record this data.

e. Ask students, "Which angle produced the greatest distance? Try to explain this result using what you know about motion."

EXPLANATION

The following explanation is intended for the teacher's information. Modify the explanation for students as required.

The catapult gun stores elastic potential energy in its spring. The catapult portion of the catapult gun is controlled by a U-shaped spring, much like a spring found in a single-hole hole punch. Energy is stored when the tray of the catapult gun is pushed down into "loading" position. A simple lever mechanism holds the loaded tray in place. When the trigger of the catapult gun is pulled, the lever moves backward, allowing the compressed U-shaped spring to spread apart. The spring transfers its energy to the loaded launching tray. As the launching tray springs upward, some of the energy is transferred to the projectile. Part of the potential energy stored in the spring is converted to the kinetic energy present in the moving projectile. The rest of the original energy is in the kinetic energy of the launching tray and is converted to thermal energy when the tray reaches its upright position and is stopped.

The kinetic energy of an object is proportional to its mass and its speed. For the same kinetic energy, an object with large mass will have a small speed, and an object with small mass will have a large speed. Thus, when the potential energy of the spring is converted into kinetic energy of the launching tray and the object, objects with larger mass will acquire less speed.

The speed of the object when it leaves the catapult gun determines how far it goes before hitting the floor. Since all objects are fired from the same height and angle, all the objects take the same time to reach the floor. Thus, those that are moving with a greater speed horizontally will travel farther in that amount of time.

The size of the air resistance force (friction with the air) on a projectile depends mostly on its speed but also to some extent on its surface area and composition. (None of the projectiles except the Styrofoam had a large surface area, so the effect of differing surface areas was minimal.) The faster the projectile travels, the larger the air resistance force is. If a light object and a heavy object having the same shape travel through the air with the same speed, the force of air resistance on them will be the same. However, the effect of that force is not the same. The force will cause the light object to slow down much more rapidly than the heavy object. This is a result of Newton's second law: the acceleration produced by a force is inversely proportional to the mass of the object. Thus, a small mass has a large acceleration. The acceleration produced by the air resistance force is in the opposite direction of the object's motion, so the object slows down. This explains why the Styrofoam, which might be expected to travel the greatest distance, in fact does not go very far.

Anomalous Data and the Subjective Aspects of Science

Part of this experiment requires dealing with anomalous data. This provides an opportunity to discuss the subjective aspects of science, an idea often overlooked in the study of science. While science is often regarded as coldly objective, it actually involves many personal decisions. Scientists must decide what they want to investigate. They must decide what methods to use to collect their data, and they must decide when sufficient data has been collected to allow them to make valid conclusions. All of these decisions can reasonably be questioned by other scientists or even by ordinary citizens as they try to decide how scientific research will impact their lives. For instance, if you read in the newspaper that a scientist has shown that living near electric power lines may cause leukemia, you may want to find out more about how the scientist came to this conclusion before selling your house and moving if you live near electric power lines.

One important area in which subjective decisions are made in science involves dealing with occasional data that just don't fit with all the rest. The National Science Education Standards state that students need to learn to think critically about evidence and that this includes deciding what evidence should be used and accounting for anomalous data. Scientists must think carefully about what might have caused these data to be different, not just throw it out thinking, "I must have made a mistake." In this experiment, the distance traveled by a very low-mass object will be short, when logically it should be the greatest distance. Students need to think about why this is so and not just ignore these data.

You might want to tell your students the story of the discovery of Neptune. More than 150 years after Newton published his theory of gravity as the explanation of what controls the orbits of the planets around the Sun, two mathematicians, Urbain Leverrier and John Couch Adams, working independently of each other at the same time applied this theory to the measured orbit of Uranus. They found discrepancies between the observed orbit and the predictions of the theory of gravity. They wondered whether this could be because another unknown planet was also exerting gravitational forces on Uranus, and used the laws of gravity to predict where this planet would have to be to make their calculations of the orbit agree with the observed orbit. They then asked astronomers to look for a planet in that location, and Neptune was discovered very near where the mathematicians predicted it would be. This was strong evidence for the correctness of Newton's theory of gravity.

Now you may ask what this has to do with anomalous data. Historians of science have discovered that another astronomer, Joseph-Jérôme Lefrançais de Lalande, also observed Neptune nearly 50 years before its official discovery. Neptune is so far away that in the telescopes available at the time it would have looked just like a star. However, except over very long periods of time, stars don't move relative to one another. Planets, on the other hand, can be seen to move relative to the stars over periods of just days. On two different occasions Lalande recorded an object that moved relative to the stars. On both occasions he decided he had made a mistake and chose not to investigate further. The new data did not fit his preconceived idea that stars do not move relative to one another. If he had instead observed this strange object further, he would now be known as the discoverer of Neptune.

The Extension

In the Extension, the projectile launched at the 45° angle should travel the greatest distance. It receives a force that propels it forward and upward. Gravity slows and stops its upward climb and then causes it to fall with increasing speed toward the ground. Since the projectile travels upward before it begins to move downward in its flight, it stays in the air longer than a projectile fired in the horizontal position. (A projectile fired in the horizontal position starts to fall, or accelerate toward the ground, as soon as it leaves the catapult gun). A projectile fired at 30° has less "upward" travel time than one fired at 45° and thus stays in the air a shorter time. The decreased time affects how long the object can maintain its forward, or horizontal, motion. A projectile fired at 60° has a greater upward traveling time than one launched at any of the other firing angles, however less of the force applied to the projectile propels it forward. The 45° angle gives the best combination of forces acting on the projectile to give it the maximum traveling distance. If using very lightweight projectiles, your students may get results that differ somewhat from the theoretical 45° maximum, because air resistance becomes important when the weight of the object is small.

CROSS-CURRICULAR INTEGRATION
Math:
- The measuring and graphing involved in this activity can easily be turned into a math lesson.

Social studies:
- Research the history of the catapult. Look particularly for ways the catapult was used as something other than a weapon.

CONTRIBUTORS

Sandy Van Natta, White Oak Middle School, Cincinnati, OH; Teaching Science with TOYS research teacher, 1996.

Pam Bauser, Moraine Meadows Elementary School, Kettering, OH; Teaching Science with TOYS, 1994–95.

Loop-the-Loop Challenge

...Students use Darda coasters to explore gravitational potential energy, kinetic energy, energy transformations, and centripetal force.

✔ Time Required

Setup   10 minutes
Performance 20–40 minutes*
Cleanup   5 minutes

* Performance time will vary with method chosen. (See the Procedure.)

A Darda car on a loop-the-loop track

✔ Key Science Topics

- gravitational potential energy
- kinetic energy
- centripetal force

✔ Student Background

Students should be familiar with gravity, kinetic energy, elastic potential energy, and the idea that energy can be transformed from one type to another. This activity introduces the concept of gravitational potential energy, so no previous knowledge of it is expected. If students have previously studied circular motion, the concept of centripetal force can be reviewed and reinforced with this lesson. If not, that concept may be omitted.

✔ National Science Education Standards

Science as Inquiry Standards:

- Abilities Necessary to Do Scientific Inquiry
  Students observe the motion of a toy car on a racetrack.
  Students use their observations as a basis for explaining why the car sometimes falls.

Physical Science Standards:

- Motions and Forces
  The force of gravity and the force of the track combine to cause the car to move in a circle, continually changing its direction of motion.

- Transfer of Energy
  Energy can be transferred to the toy car by pushing it, winding a spring, or lifting it.

✔ Additional Process Skill

- defining operationally   After developing a need for the term, students define gravitational potential energy as the energy an object has due to its position above the ground.

MATERIALS

For the Procedure
Per group, or per class if done as a demonstration or Learning Center
- pull-back car
- loop-the-loop track

Toy cars and tracks are manufactured by Darda®, Majorette®, Hot Wheels®, and others. We have found the Darda cars to be particularly durable. Parts of the activity are easier to do if you have at least 3 feet of straight track on one side of the loop.

For the Extensions
❶ Per class
- pull-back car
- track with jump

❷ All materials listed for the Procedure plus the following:
Per group, or per class if done as a Learning Center
- measuring device, such as a meterstick or tape measure

SAFETY AND DISPOSAL

No special safety or disposal procedures are required.

GETTING READY

1. Assemble at least one track to make a loop with at least 2 feet of straight track on either side.

2. If using a Learning Center or doing the Extensions, prepare index cards containing instructions. (A template for Instructions for the Learning Center Investigation is provided.)

INTRODUCING THE ACTIVITY

Pull back and release the car on a table or the floor. Pull the car back again, but this time do not release it. Ask students what kind of energy the car has now and what kind it will have after it is released. Demonstrate the car on the track several times. (You may wish to appoint a student to catch the car at the end of the track.) Vary the car's speed so that sometimes it goes around the loop and other times it does not. Tell the students they will be investigating what determines whether the car makes the loop and looking for other ways to use the toy without winding the motor.

PROCEDURE

Option A: Small Group Hands-on Investigation

Instruct each group to assemble their track in the same way yours is and then conduct the investigation and answer the questions on the Group Record Sheet (provided).

Class Discussion for Option A

When all the groups have finished their work, bring the whole class back together to discuss results.

1. Ask what seems to determine whether the car makes it around the loop. *Initial speed, initial kinetic energy, initial elastic potential energy, and how far it was pulled back are all acceptable answers.*

2. Ask whether the car speeds up or slows down as it goes up the loop. This can be difficult to notice when the car is moving rapidly. If the car is moving at speeds slow enough that it just barely makes it around, it clearly slows down as it goes up.

3. Ask where along its course the car has the most kinetic energy. Although it is hard to point to an exact spot in the motion, it is when the motor has just finished unwinding. After that the car has no source of additional kinetic energy and will start losing KE as friction converts it to heat. Some students will probably say it has the most KE as it leaves the loop. Ask them where the additional energy came from so that it could leave the loop with more KE than it entered with.

4. Ask a group to demonstrate one of the methods they found to get the car to go around the loop without winding the motor, then have a different group demonstrate a second method. While there are lots of ways to make the car move, only two are likely to get it going fast enough to make the loop: pushing it by hand and raising one end of the track to make a hill for the car to roll down.

 It may be necessary to add an extra section of track to the raised end to make the hill long enough.

5. Ask someone where the car's initial energy came from. In all cases, it is from the chemical energy in your body as you exert a force on the car (doing work) to push it, lift it, or wind the motor.

6. Discuss how the energy is stored for each method of starting the car. (This is the crucial part of this lesson.) Ask the students for the name given to the energy stored in the spring when the car is wound. *Elastic potential energy.* Ask the students whether any energy is stored when the car is pushed. *No, the input energy goes directly into kinetic energy without being stored first.* Lift the end of the track, then lift the car onto

the track. Ask the students whether any energy is stored in the car now. Some may say no, because the spring isn't wound. Others may say there must be because the car will have kinetic energy once you release it. Point out that the car has the "potential" to move when released because of its position, just as it does when the spring is wound. Explain that when you lifted the car you did work, transferring some of your energy to the car. Since the car does not immediately start moving, the energy must be stored somewhere. We call this kind of stored energy "gravitational potential energy," because it wouldn't be there in the absence of gravity. The higher up the car is, the more gravitational potential energy it has. As the car goes down the hill, this gravitational potential energy is converted to kinetic energy. The reason the car slows down when going up the loop is that some of its kinetic energy is converted back to gravitational potential energy as it rises.

7. Ask students to think about how a real roller coaster is made. (Initially, some kind of motorized system pulls the cars up a tall hill, giving the cars lots of gravitational potential energy. Once they start down the hill, the cars are not given any more energy. Energy is continually converted back and forth between kinetic and gravitational potential energy as the cars go down and up the hills. All other hills must be lower than the first, or the cars will not have enough energy to make it to the top.)

8. If your students have previously studied centripetal force, before closing the lesson return to the question of why the car falls sometimes and not others.

9. Demonstrate several unsuccessful loops using either method (pushing the car or making a hill). Ask the students, "Why did the car fall?" Make sure they notice that the car does not stop before falling, so the problem is not that it loses all its energy of motion and falls. Make the connection to the "weightlessness" of the astronauts in the space shuttle. (See the Explanation.) If the car is going fast enough that the centripetal force required is greater than the weight of the car, then it makes the loop.

 Step 9 can be omitted if students do not have previous experience with centripetal force.

Option B: Teacher-Led Demonstration/Discussion

If you can't obtain multiple sets of cars and tracks, lead the class as a whole through the activity.

Experimenting as you go, ask the students each of the questions on the Group Record Sheet in sequence. If you wish, give the students copies of the sheet on which to record the class's answers. Discuss the answers to each question as you go along, following the outline in the class discussion for Option A.

Option C: Learning Center Lesson

Another option, if you have only one track, is to allow pairs of students to complete the investigation in a Learning Center. Individual work is not practical because someone is needed to release the car and someone to catch it.

Before deciding whether to do this or the teacher-led discussion, think about how well your students work independently and whether your classroom is large enough that students working in the Learning Center will not disturb the rest of the class.

Copy the Instructions for the Learning Center Investigation, cut out the steps along the dotted lines, and tape the steps to index cards. You may also want to prepare a worksheet for students to record their answers to the questions on the cards. They can then turn this worksheet in as proof that they completed the activity.

After all students have used the Learning Center, conduct the class discussion described in Class Discussion for Option A.

EXTENSIONS

1. Some brands of tracks have jumps in them. Discuss why the car is able to cross the jump. Students could also investigate what happens when the jump is widened, what happens when one side is higher than the other, and whether it matters if the track on either side of the jump is level rather than slanted. (Student instructions for a hands-on learning center version of the investigation are provided.)

2. The pull-back cars can be used for a hands-on activity involving measurements if students pull the cars back set distances to wind them and then determine (a) whether the car makes it through the loop and (b) if the car does make the loop, how far it travels afterward. One interesting question to ask is, "Do you reach a point where further winding does not make any difference?" Also, "Does a linear relationship exist between winding distance and travel distance?" (For example, "Does pulling it back twice as far make it go twice as far?") This part could also be done without the track by running the car on the floor. (Instructions for small group or learning center investigation are provided.)

EXPLANATION

The following explanation is intended for the teacher's information. Modify the explanation for students as required.

In order to make the car move, you must give it energy. You can do this by pushing it, winding its motor, or lifting it (and the track) up. As the car begins to go up the loop, it gains gravitational potential energy and thus must lose kinetic energy (energy of motion). If the car makes it around the loop, it regains its kinetic energy on the way down, except for a small amount used up as work done to overcome friction. If the car does not make it around the loop, it still has some kinetic energy left when it falls; the reason it falls is not because it lost all its energy.

So why does the car sometimes fall? To understand why, we need to look at the two forces acting on the car as it goes around the loop—gravity and the force of the track against the wheels of the car. Together, these forces provide the centripetal force that keeps the car moving in a circle. If you plan to discuss centripetal force with your students, a detailed explanation is provided below. If you do not want to discuss centripetal force with younger students, it is acceptable to allow them to say that the speed of the car determines whether it makes it around the loop.

Centripetal Force

Just before the car enters the loop, it is moving along a straight path. If no new force were exerted on the car, it would continue along this straight path forever. For the car to change direction and start moving in a circle, a force must act on it. As the car enters the loop, a force is exerted against its wheels by the track. (See Figure 1.)

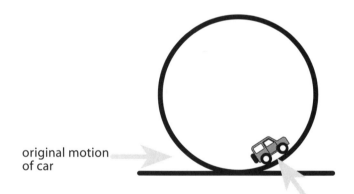

original motion of car

force of track against wheels

Figure 1: The track exerts a force that changes the direction of the car's motion.

The speed and mass of the car and the diameter of the loop determine the size of the force the track must exert to make the car change direction and start moving in a circle. The faster the car is moving, the greater the force of the track must be. As the car continues to move around the loop, it is continually changing direction and requires a continuous force towards the center of the circle (centripetal force) to make it do so.

You can visualize the role of centripetal force by imagining what would happen if, at any point in the car's path around the circle, a car-sized hole suddenly appeared in the track, right in front of the car. The car would shoot out of the hole, moving in a straight line tangent to the circle for that brief instant. (See Figure 2.) Then the car would follow a parabolic trajectory caused by the gravitational force. If no gravity acted on the car after it exited the hole, it would continue to move at the same speed and in the same direction.

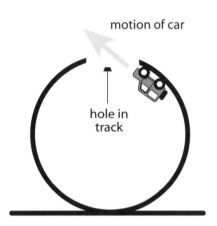

Figure 2: If a hole appeared in the track, the car would travel in a straight line out of the loop.

As the car goes around the track, two changes in its motion take place. As we just saw, the car is changing direction. It is also slowing down as some of its kinetic energy is being converted into gravitational potential energy. Both these changes must be produced by a force. The direction change requires a force towards the center of the circle (perpendicular to the track) and the speed change requires a force directed opposite the direction of the motion (tangent to the track). Where do these forces come from? The force perpendicular to the track comes from two sources: One component is the force the track exerts on the car, which is at all points perpendicular to the track (just as the force a table exerts on a book which is lying on it is perpendicular to the table). Another component of the perpendicular force comes from the force of gravity. The force of gravity always acts vertically down on the car. To understand the effect of the gravity force, we need to think of it as consisting of two parts: one perpendicular to the track (p) and one tangent to the track (t), which add together to make the actual downward force (g). (See Figure 3.) Thus, the force of gravity also provides the force tangent to the track.

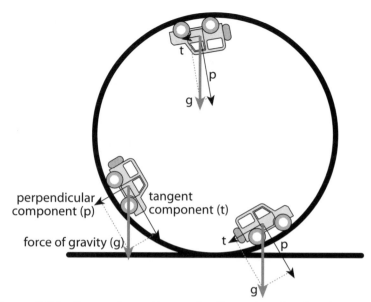

Figure 3: The force of gravity can be thought of as having two parts. The lengths of the perpendicular component (p) and the tangent component (t) are determined by the size of the rectangle for which g is a diagonal.

The relative sizes of the perpendicular and tangent parts of the gravity force change as the car moves around the circle. (See Figure 3.) Near the bottom or top of the loop the tangent part is small and the perpendicular part is large. Near the middle of the loop, the tangent part is large and the perpendicular part is small. As mentioned previously, the part of the gravity force tangent to the track is the force that changes the speed of the car. The part of the gravity force perpendicular to the track is added to the perpendicular force of the track pushing on the car, and together they provide the force needed to make the car change direction. Any force that makes something move in a circle is called a centripetal force, so this combination of perpendicular forces is often referred to as the centripetal force even though it is not a single force.

We have ignored friction between the track and the car, because it is small and would not significantly change the discussion.

Remember that for the car to move in a circle, the net perpendicular force must be inward. When the car is in the bottom half of the loop, the part of the gravity force that is perpendicular to the track points away from the center of the circle. For the net perpendicular force to be inward, the track must exert a greater force than the perpendicular portion of the gravity force. When the car is in the top half of the loop, the perpendicular gravity force points toward the center of the circle. As a result, the track force can be smaller and still result in a net perpendicular force inward.

Now we are ready to answer the question with which we began: Why does the car sometimes fall before it goes all the way around the loop? The faster the car goes, the larger the size of the centripetal force needed to change the direction of the car's motion. If the car has a high speed, then the centripetal force needed is always larger than the perpendicular part of the gravity force and the track pushes inward to provide the additional force. Thus, the car stays in the loop. If the car is moving slowly, then the centripetal force needed may be smaller than the perpendicular part of the gravity force. The track does not need to push inward. In fact, part of the gravity force is "left over"; it isn't needed to move the car in a circle. The track of course can't grab the car and pull up to cancel the extra gravity force, so the car falls.

The car's successful trip around the loop is similar to a space shuttle's orbit around the Earth. A space shuttle and astronauts are not truly weightless; the Earth still exerts gravitational force on them that is only about 10% less than it would be on Earth. When we say that the astronauts and other objects in the shuttle are weightless, we really mean that they behave as if no gravitational force was acting on them. For example, objects released in mid-air in the orbiting shuttle do not fall to the floor. This behavior is possible because the speed of the space shuttle is adjusted so that all of the gravitational force on the shuttle and its contents is always exactly the size of the centripetal force needed to make the shuttle go in a circle at that speed and radius. Thus, the shuttle does not need a track to travel around the Earth. In the activity, the car does need a track, because the car's speed can vary. In both the orbiting shuttle and the car negotiating the loop, no gravitational force is "left over" to make them fall.

The Extensions

In Extension 1 you jump the car over a gap in the track. This activity involves two-dimensional motion. As the car approaches the jump, it is moving horizontally. As soon as the car leaves the track, it continues to move horizontally and also begins to fall. Because the car is falling, the track on the landing side must be lower than the track on the first side (if the track leading up to the jump is level). Whether a car makes a particular jump depends on the size of the gap and the car's initial horizontal speed. If the car crosses the gap before it has fallen below the level of the track on the landing side, then it lands safely. If the car falls too far before reaching the landing side, it crashes into the gap.

If the track leading up to the jump is slanted slightly upward, the car leaves the track moving upward as well as horizontally. As the car moves horizontally, it will continue its upward motion, gradually slowing, and eventually beginning to fall. Because the car moves upward initially as it crosses the gap, it may not be below the starting height when it reaches the other side. So the track on the landing side may not have to be lower. Also, the initial upward motion enables the car to stay in the air longer, so the gap can be wider.

CROSS-CURRICULAR INTEGRATION

Art/Language arts:

- Having used the pull-back car in a number of ways by this time, students could make advertisements for the toy car and the track as a language arts/graphic arts/drama activity. The advertisement could be a billboard, a magazine or newspaper ad, or a TV commercial. Students should decide whether they are targeting the ad toward kids or toward teachers and parents.

FURTHER READING

Faughn, J.; Turk, J.; Turk, A. *Physical Science;* Saunders: Philadelphia, PA, 1991. (Teachers)

Kirkpatrick, L.D.; Wheeler, G.F. *Physics: A World View;* 2nd ed.; Saunders: Philadelphia, PA, 1995. (Teachers)

Zubrowski, B. *Raceways: Having Fun with Balls and Tracks;* William Morrow: New York, NY, 1985. (Students)

CONTRIBUTOR

Anita Kroger, Gifted and Talented Specialist, Cincinnati, OH; Teaching Science with TOYS, 1986–87.

HANDOUT MASTERS

Masters for the following handouts are provided:
- Group Record Sheet
- Instructions for the Learning Center Investigation

Copy as needed for classroom use.

Name _____   Date _____

Loop-the-Loop Challenge
Group Record Sheet

Run the car around the track several times, paying attention to how it behaves in the loop.

1. After having experimented with the car and track, what do you think is the most important factor in determining whether the car makes it around the loop?

    _____

2. As the car goes up the loop, does it seem to speed up or slow down?

    _____

3. When do you think the car has the most kinetic energy?

    _____

4. Find and describe two ways to get the car to go around the loop without winding the motor.

    1. _____

    _____

    2. _____

    _____

5. For all of the methods you have used to make the car move, where does the car's initial energy come from?

    _____

    _____

6. When you wind the motor, your energy is stored in the spring. Is your energy ever stored when you use the other two methods? If so, how?

    1. _____

    _____

    2. _____

    _____

Loop-the-Loop Challenge
Instructions for the Learning Center Investigation

If you are doing this activity as a learning center investigation, cut out the Loop-the-Loop instructions (steps 1–11), and mount them on index cards, one card per step. This is a good technique for presenting the instructions when you want to include information on a step that you don't want students to know at the outset. You may also want to prepare a worksheet on which the students can record their answers to the questions posed on the instruction cards.

If you are doing Extensions 1 and/or 2, also cut out the instructions labeled "Jumps" (steps 1–5) and/or "Measurement" (steps 1–4).

NOTE ABOUT STEP 7: When/if the students come to you for permission, feel comfortable telling them to go on. The next card is not an answer but a hint.

✂ -

Loop-the-Loop

❶ Wind your car and run it on the floor or a table several times.

Loop-the-Loop

❹ Why does the car sometimes go around the loop and sometimes not? What determines this? Discuss this question. Write down your answers and then share them. If you don't agree, discuss and try to come to an agreement.

- -

Loop-the-Loop

❷ Play with your car and track until you can make the car go around the loop several times.

Loop-the-Loop

❺ Run the car so that it will go around the loop again. This time watch to see where its speed is increasing and where it is decreasing. You may need to do this several times. Record your observations.

- -

Loop-the-Loop

❸ Run your car into the loop in such a way that it will not go around the loop.

Loop-the-Loop

You should have determined that the car slowed down as it went up the loop and speeded up as it came down the loop. If this was not what you saw, stop and discuss your observations with your teacher before continuing. (If your teacher is busy, record your observations, assume these speed changes occur, and see your teacher as soon as possible.) Discuss why the car slows down as it goes up. Write down the reason on your record sheet.

- -

[Continued]

Reproducible page from *Exploring Energy with **TOYS*** published by Terrific Science Press™

✂ ---

Loop-the-Loop

❼ You've been running the car around the loop by pulling it backward to wind its motor. Now find another way to make the car go around the loop without winding its motor or pushing it. This may be hard, but it can be done. After you come up with your idea, try it. DO NOT GO ON TO THE NEXT CARD UNTIL YOU COME UP WITH YOUR IDEA AND TRY IT OR UNTIL YOU GET PERMISSION TO GO ON.

Loop-the-Loop

SKIP IF YOU SOLVED THE MYSTERY OF CARD ❼.

❼a Think of a very tall roller coaster. Does this give you any ideas? Work on this a while. If this doesn't help, try the suggestion on ❼b.

Loop-the-Loop

❼b Lift up the beginning of the track.

Loop-the-Loop

❽ Measure how high you must place the beginning of the track for this method of starting the car to work. Think and talk about why this works. Where is the car's energy coming from now? How is this different from what happens when you pull the car back?

Loop-the-Loop

❾ Run the car by pulling it back, and have it run the loop unsuccessfully. Observe it carefully as it runs the track and falls and record your observations.

Loop-the-Loop

❿ Run the car by lifting the track and have it run the loop unsuccessfully.

Loop-the-Loop

⓫ Why did the car fall? Discuss this question. Did the car stop before falling? Try again if you could not tell. What does this tell you? Write your answer on the sheet.

Jumps

❶ Run the car so that it successfully makes a jump. What makes the car able to complete the jump? Discuss and record answers.

[Continued]

Jumps

❷ What happens if the jump is made wider? Try jumps of various widths and record what happens.

Measurement

❶ Lay a meterstick on the floor. Put the car down beside the meterstick and, pressing down on the back of the car, pull it back 20 centimeters. Let it go. Mark where it stops and measure the distance it traveled. Record your observations.

Jumps

❸ What happens if one side is higher than the other? Try this with various heights and record what happens.

Measurement

❷ Repeat the above, pulling the car back 30, 40, 50, and 60 centimeters. Keep increasing the pull-back distance if you'd like. Record your observations and note any patterns. Graph the results. Can you use this graph to make predictions? If so, write some predictions for distances you haven't tried and then try them out. Were you correct? Do several trials and average them. Is this answer closer to your prediction?

Jumps

❹ What happens if one side of the track is slanted? What if they're both slanted? What if they're both slanted similarly or differently? Try these adjustments and record your observations.

Measurement

❸ Now repeat step 2, running the car through the loop for each pull-back distance. Pull the car back 20 centimeters and see whether it goes through the loop. If it does so, measure how far it goes after it leaves the loop. Record your observations. Continue with distances of 30, 40, 50 centimeters, etc. As before, graph, predict, and try it out.

Jumps

❺ Discuss the reasons for the observations you made and write down your comments.

Measurement

❹ Identify some pull-back distance that will get your car through the loop. Compare the results when you use a pull-back distance of the chosen length on the track with the results of using that pull-back distance to run the car on the floor. Why do you think this happened? Write your ideas about this.

Homemade Roller Coaster

...Students make their own roller coasters and investigate the effect of track height on the distance traveled by the marble "car."

✔ Time Required

Setup	10–20	minutes
Performance	60–80	minutes*
Cleanup	5	minutes

* Two class periods

A homemade roller coaster

✔ Key Science Topics

- force
- gravity
- kinetic energy
- potential energy

✔ Student Background

Students should be familiar with the concepts of inertia, force, kinetic and potential energy, and energy conversions. (As written, the focus of this activity is energy. With little modification it could also be used in a lesson on centripetal force for older students.) The activities "Loop-the-Loop Challenge" and "What Makes It Go?" would provide students with sufficient background in energy concepts.

✔ National Science Education Standards

Science as Inquiry Standards:

- Abilities Necessary to Do Scientific Inquiry

  Students investigate how the stopping position of the marble is affected by the initial height of the track.

  Students measure the distance traveled by the marble and use the data they have collected to predict how the track should be arranged to create a particular stopping position.

  Students design roller coasters based on the knowledge they have gained.

  Students communicate their designs to fellow students, including problems encountered.

Physical Science Standards:

- Transfer of Energy

  Energy is associated with mechanical motion and can be transferred between kinetic energy and gravitational potential energy.

MATERIALS

For the Procedure
Per group
- 5- to 6-foot-long piece of foam pipe insulation (¾-inch to ⅞-inch internal diameter and thickness no greater than ⅜ inch)

These will be cut open to make two pieces of track. (Paper towel tubes and toilet paper tubes may be substituted if the foam is not available. See the Variation.) Foam insulation is available at any hardware or building materials store. However, it is seasonal and more difficult to find in the summer. Some brands of insulation are bumpy on the inside, while others are quite smooth. The smoother the inside, the easier it is for the marble to travel through a loop or barrel curve, but the distance traveled in the early part of the experiment is much larger and thus less convenient. You may want to use "bumpy" insulation for the first part and smooth insulation for the later steps involving loops.

- masking tape
- marble
- small soup or vegetable can, or other object for creating a hill in the track
- meterstick
- scissors (not needed for students if teacher pre-cuts insulation)

For the Variation
Per group
All materials listed for the Procedure except
- substitute tubes from paper towels or toilet paper for the foam pipe insulation

For the Extensions
All materials listed for the Procedure plus the following:
❶ Per group or per class
- (optional) sandpaper
- (optional) paper towels
- (optional) cooking or mineral oil

❷ Per group or per class
- large quantity of track

SAFETY AND DISPOSAL

No special safety or disposal procedures are required.

GETTING READY

If teaching students in the lower end of the suggested grade range, you may want to cut the insulation in half lengthwise to create two long U-shaped tubes for the marble track prior to class. You may want to leave one piece round but cut into shorter segments to make tunnels during the later part of the experiment.

This lesson is intended to be initially teacher-directed with the students using the Data/Observation Sheet (provided) to record measurements and answer questions, then transitioning to free exploration near the end. You should plan in advance how you will break the instructions into segments (this will vary with the age of your students). Ideally, you will give some directions, the students will do some part of the activity, the class will discuss it, then you will give the directions for the next part. For instance, you might give the instructions for steps 2–5 below; then, when the students have completed this, discuss the results including the question in step 6; and then give the instructions for step 7.

INTRODUCING THE ACTIVITY

Ask, "Have you ever been on a roller coaster?" Ask someone to describe what it looked like. Ask why roller coasters almost always start with a big hill.

Ask students if they have ever been on a roller coaster ride that turned them completely upside down. (If you expect that many of your students have not done this, you might want to show a short video clip of a roller coaster ride.) Ask them how they felt when turning upside down. Ask why they did not fall out of their seats. Do they think they would have fallen out of their seats if the safety bar was not in place? If you have done the activity "Loop-the-Loop Challenge," you can remind students of what they learned by running a toy car through a track loop and asking them to explain what determines whether the car goes around the loop and what energy conversions take place during the trip.

An alternative introduction with intermediate students is to read *Harriet and the Roller Coaster,* by Nancy Carlson (Trumpet Club, ISBN 0440845912), with the class. In this story, Harriet accepts her friend George's challenge to ride the frightening roller coaster and finds out that she is the brave one.

PROCEDURE

> *Give students instructions for a few steps at a time, and discuss results prior to giving them additional instructions. Feel free to group the steps as you consider most manageable. Students can either answer questions on the Data/Observation Sheet, or they can just discuss orally with the class.*

1. Divide the students into groups of four and assign jobs. Each group will need a Marble Roller, Measurer and Marble Retriever, Recorder, and Track Holder.

2. Have students cut the insulation in half lengthwise if you have not done so already. Have them use masking tape to join two pieces of foam insulation together end to end to make one continuous track 10–12 feet long. If possible, keep all the tape on the underside of the track to prevent changes in friction as the marble passes over the foam track.

3. Have one student per group hold one end of the track against a wall about 15 cm above floor level. (See Figure 1.) The concave side of the track should face upwards. Try to keep the incline as straight as possible (no sagging) as it slopes toward the floor. The track should form a short inclined plane to the floor and then continue flat against the floor's surface. The exact angle of the track is not important; just make sure the angle is not so steep that the track has a sharp kink at the bottom.

Figure 1: Hold one end of the track against the wall, about 15 cm above the floor.

4. Students should hold the marble so that it lines up with the top of the track against the wall. This can be done by placing a pencil across the track just beneath the bottom of the marble. (The marble can be released in step 5 by simply pulling the pencil out of the way to allow the marble to roll down the groove in the track.)

Using a pencil ensures that students do not add a push to the marble as they release it.

5. Have students guess where the marble will stop after being released from the top of the incline. Then have them release the marble.

6. Have students measure how far away from the wall the marble stops if it is still on the track. If the marble continues rolling off the track and onto the floor, the distance can be recorded as "off the track."

7. Ask students to brainstorm a list of factors that might have affected how far the marble rolled.

8. Ask students to investigate the effect of changing the height of the starting point, performing several (3–4) trials at each height and recording data on the Data/Observation Sheet. Remind students to keep the incline straight as the track slopes toward the floor rather than have it go vertically downward, then curve outward.

9. Ask students, "Did the stopping point of the marble change when the height of the track changed? If so, did doubling the track height make the marble travel twice as far?"

10. Ask students to explain any difference between the results of the trials in terms of energy.

11. Ask students to pick a location along the track and mark it with a small piece of masking tape. Have them use their data to choose an initial height that will cause the marble to stop at the chosen point (without any barrier being placed at this point). Have them release the marble and see if their prediction was correct.

12. Have students choose an initial height and keep it constant. Have them change the shape of the track in various ways. Students can make a low hill (a hill shorter than the height of the starting point) in the track by running the track over a small can, or they can put a curve in the track using one or two students to hold the curve in place.

13. Ask students whether any of these changes make a difference in where the marble stops. If the marble has the same initial amount of energy in each trial, should the shape of the track make a difference? (Theoretically, there should not be any change. However, students may have noticed a change; this change results from the fact that when the track shape changes, the track cups the marble differently, changing the amount of the track surface in contact with the marble.)

14. Have students experiment with making a loop in the track. They should change the shape, size, and position of the loop until the marble is able to make it completely around the loop and continue onward on the track. (See Figure 2.) They may try round loops; tall, thin loops; and short, fat loops. They can place the loop near the beginning, middle, and end of the track. Ask students to determine whether the shape or location makes any difference.

Figure 2: Have students experiment with different sizes and shapes of loops.

15. Have students try a barrel roll (a spiral loop that has a more horizontal orientation than the loop in step 14, as shown in Figure 3). Was the marble able to negotiate this shape?

side view   top view

Figure 3: Set up a barrel roll in the track.

16. Challenge groups to combine two or more marble tracks and design a big roller coaster with as many hills and curves as they wish. Have them test and revise until they produce a successful design. Then have them draw their design to scale.

17. Allow each group to demonstrate its final design and explain the process used to develop the final design. Ask them to explain how they overcame any difficulties.

EXPLANATION

The following explanation is intended for the teacher's information. Modify the explanation for students as required.

The marble gains energy when it is lifted to the top of the track. It stores this energy as gravitational potential energy until it is allowed to roll down the track. Once the marble is in motion, gravity accelerates it downward, and its potential energy is converted into kinetic energy (KE, also called energy of motion). The marble continues to roll until the force of friction stops it. The greater the initial starting height of the marble, the greater the energy stored in the marble, and the farther the marble rolls. The marble stops when all of its KE has been converted to thermal energy due to friction between the marble and the track. In general, doubling the height should more or less double the distance traveled.

When a small hill is placed in the track, the marble loses kinetic energy traveling up the hill but gains gravitational potential energy. As the ball starts back down the hill, the process is reversed. Since the marble gains one type of energy while losing another, the distance traveled should not differ significantly from the distance traveled on a flat track with the same initial height. The marble should be able to travel over hills as long as the hills do not exceed the initial height of the marble and friction between the marble and the track is not too great. Also, hills near the end must be lower than hills near the beginning.

Curves should not make a significant difference in the distance a marble travels. Only if the curve is slanted so that more of the marble makes contact with the bottom and sides of the track will friction increase enough to make a difference in the distance the marble travels.

The same trade-off of potential and kinetic energy discussed with the hill also occurs in a loop. As with hills, if the top of the loop is above the initial starting height, the marble will not have enough kinetic energy to make it to the top of the loop. However, the marble may not go around a loop that is the same height as a hill that it can make it over. With the hill, as long as the marble has any kinetic energy left at the top, it will keep going on over the hill. However, in the case of the loop, a certain minimum speed is required at the top of the loop.

The marble will continue rolling in a straight line until something exerts a force on it. As the marble enters the loop, the track exerts a force on it to make it travel around the loop. The speed and the mass of the marble and the diameter of the loop determine the size of the force the track must exert on the marble to keep it moving in a circular path. As the marble continues moving through the loop, gravity and the force of the track against the marble are combined to make up the centripetal force that keeps the marble moving in a circle. The slower the marble is traveling, the less force the track must exert. If the marble slows down enough that the track doesn't have to exert a force because gravity by itself is enough to provide the centripetal force, the marble will fall.

The marble should be able to traverse loops that are not too high. However, the shape of the loop does affect the result. Loops with gradual curves are more likely to be successful than loops with sharp turns. The location of the loop is important only in that during the trip the marble is gradually losing kinetic energy due to friction. Thus, the farther from the beginning point the loop occurs, the less kinetic energy will be available to be converted into potential energy as the marble goes up the loop. As a result, loops near the end must be lower than loops near the beginning.

Real roller coasters use a shape called a Clothoid loop, which starts out with a large radius and then gradually tightens into a smaller circle. This shape keeps riders from experiencing a very large force when they first enter the loop. (See the Explanation in "Loop-the-Loop Challenge" for more detail on how the force of the track and gravity work together to keep the marble in the loop.)

VARIATION

If foam insulation is unavailable, tubes from paper towels or toilet paper can be cut in half and taped together to make a track for the marble. If the tube ends are cut at an angle, they can be formed into curves. The resulting track is not as smooth as the foam track but will work in this activity.

EXTENSIONS

1. Ask students what can be done to slow down or increase the speed of the marble. Have students investigate the effect of lining the track with masking tape, paper towels, or sandpaper. Also, a small amount of cooking or mineral oil may be swabbed inside the track to increase the speed of the marble by decreasing friction. Allow students to brainstorm other ways to change the speed of the marble along the track. If possible, allow students to test their ideas.

2. You may wish to challenge students to find a way to keep the marble rolling continuously for some specified length of time. (You need to have a large quantity of track on hand in order for students to test their hypotheses.)

3. If the weather is appropriate for outdoor work, challenge students to make a roller coaster at home from materials they already have. One example might be to fold a garden hose in half and use the trough formed between the two halves as a track for a marble. Have students report back to the class on what they used to make their roller coaster.

CROSS-CURRICULAR INTEGRATION

Art:

- Have students translate their homemade roller coaster designs into what a real roller coaster might look like if it were configured the same way. The students could develop advertising posters for an amusement park announcing their "new" roller coaster.

Language arts:

- Students can be asked to write about their experiences on amusement park rides. Ask them to describe a ride, then identify where energy is being gained or lost, where inertia tends to keep them moving, and where they feel the effects of centripetal force and/or gravity. If any of your students have not experienced amusement park rides, you could have them write about how it might feel to be the marble in their experiment.

Math:
- This activity can easily be adapted to a math/graphing activity. Students can plot the height of the track versus the distance the marble travels or compare the marble mass to the distance traveled by marbles of different masses.

Social studies:
- Have students look up the locations of the highest, fastest, and oldest roller coasters and locate them on a map. Make a class timeline of important events in the history of roller coasters.

REFERENCE

McKean, P.B. "Roller Derby," *SuperScience Blue.* 1994, *6*(1), 20–21.

FURTHER READING

Wiese, J. *Roller Coaster Science;* John Wiley & Sons: New York, NY, 1994. (Students)

CONTRIBUTORS

Sandy Van Natta, White Oak Middle School, Cincinnati, OH; Teaching Science with TOYS Research Teacher, 1996.

Elizabeth Billittier, St. Patrick's, Oneida, NY; Teaching Science with TOYS, 1996.

Elisia Hayes, Shade Elementary School, West Carrollton, OH; Teaching Science with TOYS, 1996–1997.

Mary Ellen Jablonsky, Camden Middle School, Camden, NY; Teaching Science with TOYS, 1996.

Staci Sabato, Delshire Elementary School, Cincinnati, OH; Teaching Science with TOYS, 1996–1997.

HANDOUT MASTER

A master for the following handout is provided:
- Data/Observation Sheet

Copy as needed for classroom use.

Name _____ Date _____

Homemade Roller Coaster
Data/Observation Sheet

1. How far did the marble roll the first time you released it on the track? If the marble rolls off the track and onto the floor, record the distance as "off the track."

   _____

2. Why do you think the marble stopped at this position?

   _____

3. Record the data from your trials below.

	Height of the incline (cm)	Distance the marble traveled (cm)				Average distance marble traveled (cm)
		trial 1	trial 2	trial 3	trial 4	
Incline 1						
Incline 2						
Incline 3						
Incline 4						
Incline 5						

4. Does the stopping point of the marble change as the height of the incline changes?

   _____

5. Does doubling the height of the incline make the marble travel twice as far?

   _____

Reproducible page from *Exploring Energy with* **TOYS** published by Terrific Science Press™

6. Explain any changes you observed in potential and kinetic energy.

_____

_____

_____

7. What did you have to do in order to get your marble to stop at a chosen point?

_____

_____

8. Do hills or curves in your track cause the marble to stop at a different position? Do you think hills and curves should make any difference? Why or why not?

_____

_____

9. Draw the types of loops you put into your track. Under each one, tell whether the marble made it all the way around the loop or fell to the ground.

10. What did you have to do to make the marble complete a barrel roll?

_____

11. Draw the final design of your combined roller coaster below or on another sheet of paper.

Bounceability

...Students experiment with bouncing balls to explore the transformation of other forms of energy into thermal energy.

✔ Time Required

Setup   5 minutes
Performance   80 minutes
Cleanup   5 minutes

✔ Key Science Topics

- elastic potential energy
- gravitational potential energy
- kinetic energy

✔ Student Background

This activity should not be used as an introductory lesson on energy. Students should already be familiar with kinetic energy, elastic potential energy (springs and rubber bands), gravitational potential energy, and heat. They should have investigated other stored energy toys in activities such as "What Makes It Go?" and "Loop-the-Loop Challenge."

Bouncing balls

✔ National Science Education Standards

Science as Inquiry Standards:

- Abilities Necessary to Do Scientific Inquiry

  Students identify and control variables.

  Students design an investigation and communicate their experimental methods to other students who critique their design.

  Students carry out the investigation, collecting, graphing, and interpreting the data.

Physical Science Standards:

- Transfer of Energy

  Energy is a property of systems and is associated with mechanical motion, position, heat, and sound.

MATERIALS

For Getting Ready
- (optional) assorted colors of construction paper

For Introducing the Activity
- spring-up toy

For the Procedure
Per class
- 1 each of a variety of balls such as tennis, golf, rubber, Styrofoam™, hi-bounce, metal, clay, racquetball, basketball, and rubber playground ball
- tape
- variety of surfaces to bounce the balls on, such as carpet, grass, flooring tiles, ceiling tiles, metal plate, cardboard, cork, foam pad, and Styrofoam

 The exact number of balls and surfaces is not important, but you should have at least as many balls as groups.

Per group of 3–5 students
- meterstick or long strips of paper
- chart paper
- colored markers

For Variations and Extensions
❶ All materials listed for the Procedure plus the following:
- access to a refrigerator or radiator (or other heater)

❷ Per class
- Smart Ball™ (by Applied Elastomerics), Splat Ball, or Jelly Ball

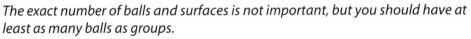 *These are all basically the same toy by different manufacturers.*

SAFETY AND DISPOSAL

No special safety or disposal procedures are required.

GETTING READY

1. (optional) Ask students to bring balls from home.

2. Put numbered lists of the balls and surfaces that will be used on the board.

3. For younger students, determining how high the ball bounces in centimeters may be difficult. If this is the case, mark the metersticks in 10-cm sections and wrap each section in a different color of construction paper. The students can then record each ball's bounce height as "blue," "yellow," or some other color, depending upon which colored section it bounces into.

INTRODUCING THE ACTIVITY

Remind the students that they have already investigated several "stored-energy" toys. Review the concepts of kinetic and potential energy using one or more of these toys. For example, demonstrate a spring-up toy and have the students describe the energy transformations that take place.

Present the following ideas to your students: "A ball would seem to be the simplest possible stored-energy toy; however, we did not begin with it because it is really a lot more complicated than you might think. Today we want to investigate the bouncing of balls to try to determine why some are better bouncers than others."

PROCEDURE

Part A: Discussing the Variables

1. Show the students the balls they will be testing. Ask them to think about which one will be the best bouncer, and then take a vote. Select a few students who voted for different balls. Ask each of these students to explain to the class his or her reasons for selecting a particular ball.

2. Ask students what variables might affect how high a ball bounces. Some ideas may have come out as students explained their choices in step 1. If so, write these on the board, then encourage the students to brainstorm other ideas. Possible variables include the material the ball is made of, its temperature, whether it is dropped or thrown, the height from which it is dropped, and the surface that it hits. At this point, do not try to limit the students to ideas that you think are "reasonable."

Part B: Planning the Experiment

 If your class does not have prior experience with planning experiments, you could develop the plan through a whole group discussion, rather than through small group work.

1. Tell the students that today you are going to investigate two of the variables that affect how high a ball bounces: the material the ball is made of and the surface on which it is dropped. Tell the students that the first step is to design an experiment that allows them to test both of these variables.

 Another option is to have students design two experiments. In one experiment, test how high one ball bounces on different types of surfaces, and in the other experiment, test how high different types of balls bounce on one surface. Combining the experiments, as described in Part C, is simply a method that may save time.

2. Divide the class into groups of three to five students. Ask each group to design an experiment to determine which surface makes each ball bounce highest. Depending on their prior experience, you may want to remind them to control variables.

3. Once all groups have completed their plans, select one group to explain its plan to the class. Have the rest of the class critique it. Is it a good plan? Have they forgotten to control one of the variables?

➤ *A reasonable experimental plan is to drop each ball from a standard height (such as 1 m) measured from a standard position on the ball (for example, the bottom), and then "eyeball" the rebound height when the ball bounces. Whatever dropping height and measuring position on the ball you choose, make sure that all students are consistent. Students may want to make a practice drop to determine which section of the meterstick they need to watch carefully, then make their measurement on the next drop.*

4. After one good plan is established, ask whether any other group has a different idea. If so, have them explain their plan to the class as the previous group did.

5. If several different plans are proposed, explain to the class that they must choose one plan for everyone to follow so that they can then compare data afterwards.

Part C: Conducting the Experiment

➤ *Students should conduct the experiment in their small groups. Each group will need a Ball Dropper, an Initial Height Checker, and one or more Rebound Height Checkers to perform the following jobs:*
- *Ball Dropper—to hold the meterstick (if necessary) and drop the ball,*
- *Initial Height Checker—to stand back and make sure the initial height is right, and*
- *Rebound Height Checker(s)—to note the rebound height.*

One way to collect a large quantity of data is as follows (this experiment tests both variables and may save time):

1. Set up the various surfaces to be tested at different locations around the room.

2. If sufficient wall space is available, tape metersticks to the wall, or tape lengths of paper to the wall so that the Rebound Checkers can mark the paper and measure the height.

3. Assign each group a different kind of ball (or two balls if you have a sufficient number of different ones).

4. Have the groups rotate through the locations with different surfaces, testing their ball(s) at each stop.

5. Instruct each group to make a bar graph of their data on chart paper large enough for the whole class to see. They should have a bar for each

surface, and the height of each bar should show the height achieved by the bouncing ball on that surface. To facilitate comparisons, each group should use a vertical scale of 0 to 100 cm. Have groups post their graphs on the wall. For examples, see Sample Graphs. These graphs are for your reference only; students should make their own graphs.

Make sure all groups put the surfaces on the chart in the same order to facilitate comparisons. Color-coding can be used to increase readability. For instance, all bars representing carpet might be green and all bars representing Styrofoam yellow. Colored markers can be used or strips of construction paper cut in advance which can then be trimmed to the appropriate length and taped on the chart.

SAMPLE GRAPHS

Figure 1: Rebound Heights of a Tennis Ball

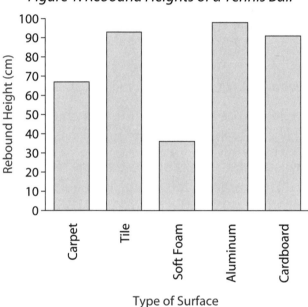

Figure 2: Rebound Heights of a Steel Ball

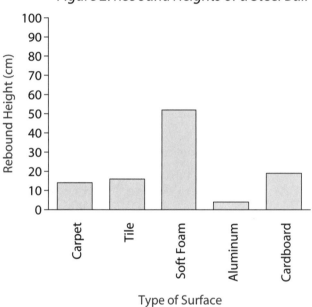

Figure 3: Rebound Heights on Soft Foam

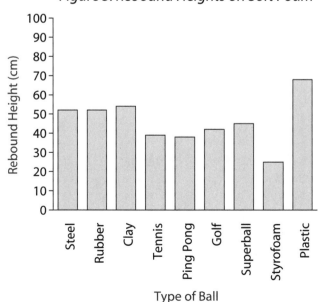

Figure 4: Rebound Heights on Cork

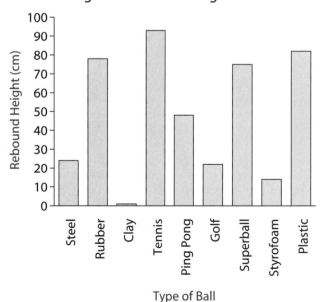

CLASS DISCUSSION

Explain to students that they have clearly demonstrated that not all balls bounce equally and that the drop surface does indeed have an effect on how high the ball bounces. Either ask the students to discuss the energy transformations that take place as the ball falls and then bounces back or ask them specific questions such as, "What kind of energy does the ball have before you drop it, while it is falling, and just before it hits the floor?" "Why doesn't the ball come back to its original position? Did it lose energy?" (See the Explanation.)

Now help students apply these ideas to the data on the graphs. Which ball is the best bouncer overall? Other questions to consider could include the following: What surface results in the highest rebound? What surface results in the lowest rebound? Why? Why does the answer depend on the type of ball?

Some of the following observations may emerge from your discussion:

- Students may note that the hi-bounce ball is rarely the best bouncer on any surface but is reasonably consistent from surface to surface.

- Ask students to look at how each of the balls bounced on the foam and the metal. In comparison with the other graphs, students should observe little variation from one ball to another on the foam. You can explain that the foam deforms easily, so the stiffness of the ball doesn't matter much. Students should observe the biggest variation on the metal plate, because it deforms very little.

- Students may observe that the clay bounces significantly only on the foam and does not bounce at all on the metal and floor tile. The clay loses its kinetic energy in a permanent shape change when it hits hard surfaces, so it can't bounce back up.

Continue the discussion as long as the students are interested in thinking about the reasons why different kinds of balls bounced differently on the test surfaces.

EXPLANATION

The following explanation is intended for the teacher's information. Modify the explanation for students as required.

Before the ball is dropped, it has an amount of gravitational potential energy that depends on its height above the floor. As the ball falls, this GPE is gradually turned into kinetic energy. The greater the ball's speed as it falls, the greater the KE. On the way back up, the KE is being turned back into GPE. At the top of the bounce, the ball's energy is once again all GPE. Since the ball does not bounce back to its starting height, it has less GPE now than it did before it was dropped.

Since energy cannot be "lost," the missing energy must be converted into some other form of energy. The energy transfer happens during the interaction between the ball and the surface. Part of the energy goes into the sound waves that are produced when the ball hits the surface. Part of the energy is converted to thermal energy (heat) in the ball and in the surface. You can observe this if you drop a ball of clay several times in rapid succession. The clay will begin to feel warmer.

What determines how high different balls bounce on the same surface? Much of the difference is a result of how much the balls deform and, even more importantly, how fast they return to their original form. During the bounce, the shape of the ball changes. This shape change takes energy, just as stretching a rubber band does. Flattening the ball is similar to compressing a spring. You get the energy of compression (elastic potential energy) back as the shape goes back to normal. The clay doesn't bounce well because it stays deformed. If the ball is still partially deformed after it leaves the floor (or other surface), the energy that was stored in that deformation does not return to kinetic energy of the ball even though the ball does later return to its original shape. The best bouncers are likely to be the ones shown in Figure 5.

Figure 5: Best Bouncers

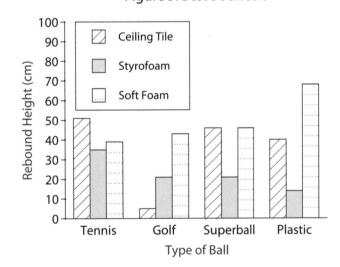

What determines how high the balls bounce on different surfaces? During the bounce, both the shape of the ball and the shape of the drop surface are deformed. The height of the bounce is determined by how much energy of compression is returned as the shape of both the ball and the surface go back to normal. Each ball type and surface type interact differently, producing a unique result. Even so, some surfaces produce fairly consistent results with all types of balls. For example, all the balls bounce on the foam, because the foam deforms rather than the ball, acting much like a trampoline. In contrast, if the surface stays deformed, as the Styrofoam may, then the energy that went into causing the deformation does not return to the ball.

EXTENSIONS

1. Test the prediction that temperature affects bouncing by cooling several balls in the refrigerator overnight or heating them by a radiator or other heater. Measure how high the heated or cooled balls bounce and compare this data with the "room-temperature" measurement.

2. Demonstrate a Splat Ball, Smart Ball™, or Jelly Ball, all of which are made of a soft polymer. If thrown at the floor, these spread out almost completely flat, then slowly reform. This is a very visible example of the deformation of the ball when it hits the floor.

 If your ball stops working properly, wash it with soap.

ASSESSMENT

This assessment can be done either individually (written) or as a group (oral). If done individually, you may want to make it a homework assignment so students won't feel pressured by time. If done as a group assessment, explain that all members of the group are responsible for the "answer," as any one of them could be asked by the teacher to respond. Groups should be given at least 10 minutes to discuss their answer. Then you can circulate among the groups, calling on one student to answer for the group. Groups should be given an additional task to work on while waiting their turn. Alternatively, you could generate a few more variations on the questions, give each group a unique question, and have oral answers presented to the whole class.

Have students apply what they have learned to answer one of the following questions:

1. How would playing basketball on a carpeted court be different from playing on a regular wooden court? Or, how is playing basketball indoors on a wooden court different from playing outdoors on a paved court?

2. You and your friends want to play baseball, but you can't find the ball. Your little brother comes up with a tennis ball. You decide to try it. How will your game be different?

3. Some tennis tournaments are played on grass. How do you think those games are different from games played on a clay court?

CROSS-CURRICULAR INTEGRATION

Language arts:
- Read and discuss the book *The Pinballs,* by Betsy Byars (Scott Foresman, ISBN 0673801365). In the book, one of the foster children describes herself and the others as pinballs—they bounce from here to there, wholly dependent on those around them.
- Have students imagine what it would be like to be a certain type of ball and explain in writing how they would react to different surfaces and how the surfaces would affect their bouncing.
- Have students invent and describe a brand-new game with a ball on a surface that is not normally used with that type of ball.

Math:
- Use this activity as a math activity, focusing on measuring, averaging, and graphing, with science as the secondary focus.

Physical education:
- Discuss why different kinds of balls are used in different sports.

FURTHER READING

Doherty, P. "That's the Way the Ball Bounces," *Exploratorium Quarterly.* Fall 1991. (Teachers)

Haduch, B.; Williams, D. *Balls: The Book with Bounce;* Planet Dexter, Ed.; Addison-Wesley: Reading, MA, 1996.

Kaplan, N. "But That's Not Fair," *Science and Children.* 1994, *21*(7), 24–25. (Teachers)

Kirkpatrick, L.D.; Wheeler, G.F. *Physics: A World View,* 2nd ed.; Saunders: Philadelphia, PA, 1995. (Teachers)

Schools Council. *Science from Toys;* Macdonald Educational: New York, NY, 1972; Chapter 7. (Teachers)

Stone, H.; Siegel, B. *Have a Ball;* Prentice-Hall: Englewood Cliffs, NJ, 1969. (Students)

CONTRIBUTORS

Anita Kroger, Gifted and Talented Specialist, Cincinnati, OH; Teaching Science with TOYS, 1986–87.

Bruce Peters, Intermediate Coordinator/Science Specialist, Middletown, OH; Teaching Science with TOYS, 1986–1987.

The Energy Transformation Game

...Students are introduced to some new forms of energy and use a modified "Guess Who?" game to reinforce the concept of energy conversion.

✔ Time Required

Setup   30 minutes (only first time games are used)
Performance 30–40 minutes
Cleanup   5 minutes

"Guess Who?" game with Energy Transformation game cards

4 Key Science Topics

- energy conversions
- Law of Conservation of Energy
- types of energy

✔ Student Background

Students should be familiar with kinetic and potential energy and should have previous experience discussing how energy is converted from one form to another. Also, they should be able to track the conversion of energy in specific devices.

✔ National Science Education Standards

Physical Science Standards:

- Transfer of Energy

  Energy has many different forms, and many devices transform energy between these various forms: Electric circuits transform electrical energy into heat, light, sound, and mechanical energy; the sun transforms nuclear energy into light and radiant energy; and chemical reactions transform chemical energy into other forms of energy.

✔ Additional Process Skills

- communicating   Students work in groups, question each other, and arrive at a consensus.

- classifying   Students categorize various devices depending on the energy they use and produce.

MATERIALS

For the Procedure
Per group of 2 or 4 students:
- "Guess Who?®" game by Milton Bradley

This game is available at many toy stores. Some of your students may have the game at home and may be able to bring it to school for a few days. The travel version of this game is more cost-effective to buy, and the card holders are smaller for easy storage. The "Guess Who?" game could be multi-purpose and used at different grade levels for other classification activities, so the purchase of 8–10 games may be worthwhile. This game can be played without purchasing the commercial version. In this case, when photocopying Game Card masters, enlarge them to the size of regular playing cards. Have students arrange the cards face up on the table, turning cards over as they are eliminated.

- 2 sets of "Energy Transformation" Game Cards on 2 different colors of paper (preferably card stock)
- set of "Energy Transformation" Mystery Cards on a third color of paper (preferably card stock)
- 8–10 copies of Clue Check-Off Sheets on white paper (4 may be copied per page)

SAFETY AND DISPOSAL

No special safety or disposal procedures are required.

GETTING READY

If necessary, read the instructions to familiarize yourself with the game. Decide which cards to use depending on the grade level and ability of the students. For younger students you may want to use only cards for devices that represent a single energy conversion such as E→L rather than a double energy conversion such as E→L, H. You can also begin with a few easy cards and add harder ones after the students have practiced.

Using two different colors of paper (preferably card stock), copy the Game Cards for Full-Size Game or Game Cards for Travel Game (provided), cut them out, and insert them in the plastic holders on the game board. Using a third color of paper, copy the Mystery Cards (provided). If necessary, remove any cards you do not wish your students to use from all three of these sets. Copy the Clue Check-Off Sheets (provided) on white paper. Shuffle the Mystery Cards and place the deck face down between the two game boards. Once this is prepared, the setup time for the activity is minimal.

Both sets of Game Card masters contain pictures of 24 items, corresponding to the 24 holders of the full-size game. The travel game has only 20 holders, so you will have to decide which items to leave out. Blanks for making your own versions of the game are also provided. (See Cross-Curricular Integration.)

INTRODUCING THE ACTIVITY

You may wish to have the students play the "Guess Who?" game as is before inserting the energy cards for the activity. Explain the rules of the game and that all questions must be answered by a "yes" or a "no." If students are older or are familiar with the game, this should not be necessary.

Review all of the types of energy that are depicted on the cards you have chosen to use. You will need to introduce any types of energy to which students have not already been introduced. (See Pedagogical Strategies for a discussion of how to introduce these forms of energy.) For younger students, you may need to go over what all of the pictures represent. Be sure to tell students that the devices may use more than one kind of energy and may produce more than one kind of energy.

PROCEDURE

1. Divide students into groups of two or four. (See Variation 1 for an alternate mode of play with groups of three or five students.) Have students play the game according to the following rules:

   a. Mystery Cards should be shuffled and placed in a stack face-down between the teams. Each team draws a Mystery Card from the deck and puts it in the Mystery Card slot in the front of the game board.

   b. The game begins with all card holders up. Teams take turns asking a "yes or no" question about the energy used by the mystery device or the form to which the device converts energy. After each question and answer, the asking team confers and keeps track of clues on their Clue Check-Off Sheet by crossing out energy forms as they are eliminated. They flip down the holders of the devices that they wish to eliminate.

   c. Play continues until one of the teams wants to guess the name of the mystery device during their turn; the team guesses the name instead of asking a question. If they guess correctly, they place a peg in the hole on the game board. If they guess incorrectly, they skip a turn (the other team is able to ask two questions in a row).

 In the original "Guess Who?" game, if one team guesses incorrectly, the other team wins the game. If students played the original game, point out that in the energy game a wrong guess will cause the team to lose its next turn, but the game will not end with a wrong guess.

d. Each time a peg is awarded, a new Mystery Card is chosen by each team, and all the card holders are flipped up again. This is considered the beginning of a new game. A new Clue Check-Off Sheet is used for each game in the match. The match is won when one team fills all five holes on their game board with pegs.

This is similar to the championship round in the "Guess Who?" instructions.

2. While students are playing, circulate around the room to make sure students are asking questions using the energy concepts and to make suggestions.

3. After all teams have completed at least one match, ask someone to summarize the point of the activity. *Many devices transform one kind of energy into another.* Be prepared with a few other examples to challenge your students. For example, what conversions take place in a telephone? How are the conversions in a cellular phone different from conversions in a regular household phone?

EXPLANATION

The following explanation is intended for the teacher's information. Modify the explanation for students as required.

Energy has many forms. Energy is constantly being converted from one form into another. However, the total amount of energy in a given system remains the same. This concept is referred to as the Law of Conservation of Energy.

For this activity, eight forms of energy are identified and should be familiar to students. A basic review of these terms may be necessary: For more explanation of these forms of energy, see the Content Review section of this book.

- mechanical energy—the sum of an object's kinetic (or motion) energy, its gravitational potential energy, and its elastic potential energy. On the Mystery Cards, "M" represents "mechanical energy."

- heat energy—kinetic energy associated with the random motions of particles that make up an object, more correctly called thermal energy. On the Mystery Cards, "H" represents "heat energy."

- chemical energy—potential energy associated with the arrangement of charged particles in atoms and molecules. This energy can be transformed during chemical changes when bonds between atoms are broken, made, or rearranged. On the Mystery Cards, "C" represents "chemical energy."

- nuclear energy—energy possessed by the nucleus of atoms as a result of the strong force that holds the nucleus together. This energy can be converted into kinetic energy in processes such as fission and fusion. On the Mystery Cards, "N" represents "nuclear energy."

- electrical energy—a form of potential energy that charged particles may have due to the electrical forces exerted on them. On the Mystery Cards, "E" represents "electrical energy."

- sound energy—the kinetic energy of moving molecules that carry sound waves. On the Mystery Cards, "S" represents "sound energy."

- light energy—energy carried by visible light. On the Mystery Cards, "L" represents "light energy."

- radiant energy—energy carried by waves similar to light but invisible, such as microwaves, radio waves, infrared rays, and ultraviolet rays. On the Mystery Cards, "R" represents "radiant energy."

The term "radiant energy" is often used to include all energy carried by electromagnetic waves, including visible light. We have listed light energy as a separate form of energy, because radiant energy may be difficult for younger students to understand. Because these terms are separated, only five cards involve radiant energy, and these can be omitted for younger students.

VARIATIONS

1. The game can be played with an odd number of players (3 or 5) if one player is the moderator. The moderator chooses a Mystery Card and answers questions from both teams. The teams take turns asking questions and marking their clues accordingly. The first team to guess the mystery device correctly is awarded a peg. This game works well as a cooperative learning activity. The moderator becomes the judge and umpire, as much discussion can take place. This game involves critical thinking and decision-making, and the inclusion of a moderator helps to develop leadership and cooperation skills.

2. Instead of having teams play against each other, have groups play simultaneously, with the teacher answering the questions. (This requires one board per pair of students, so a class of 24 would need only six games.) The first group to guess the energy device correctly receives points, bonus, or a small prize, while groups that guess incorrectly are out of the round. This method works well with students who may not be very competent at answering each other's questions or with groups that need more control or supervision. It offers a more structured environment and more opportunity for the teacher to discuss the topic as the game is played.

CROSS-CURRICULAR INTEGRATION

General science:
- This game provides unlimited possibilities for classification activities, especially in science. Use the same format as the Procedure and make cards related to plants, animals, elements, important scientists, etc. This is an excellent activity for the discussion of classification and dichotomous keys.

Language arts:
- Use the game with cards that represent parts of speech such as nouns, verbs, adjectives, adverbs, conjunctions, and prepositions.

Life science:
- Burn a Planters® Cheez Ball to show that it contains stored energy. Apply this to the way the body converts the stored energy in food into kinetic energy as we move.

Social studies:
- Use the game with cards that represent states of the union, presidents of the U.S., geographical locations, or other categories relevant to the topic of study.

FURTHER READING

Exploring Energy; Scholastic: New York, NY, 1995.

CONTRIBUTOR

Pam Addison, Norwood High School, Norwood, Ohio: Teaching Science with TOYS District Leader, 1996.

HANDOUT MASTERS

Masters for the following handouts are provided:
- Game Cards for Full-Size Game
- Game Cards for Travel-Size Game
- Mystery Cards for Full-Size Game
- Mystery Cards for Travel-Size Game
- Clue Check-Off Sheets
- Blank Cards for Full-Size Game
- Blank Cards for Travel-Size Game

Copy as needed for classroom use.

The Energy Transformation Game
Game Cards for Full-Size Game

television	person	plant
siren	firecracker	bicycle
satellite dish	electric guitar	lightstick
backhoe	candle	iron
solar calculator	magnifying glass	mixer
microwave oven	hot-air balloon	light bulb
windmill	nuclear power plant	piano
sun	tanning bed	battery

Game Cards for Travel-Size Game

sun	windmill	microwave oven	solar calculator	backhoe	satellite dish	siren	television
tanning bed	nuclear power plant	hot-air balloon	magnifying glass	candle	electric guitar	firecracker	person
battery	piano	light bulb	mixer	iron	lightstick	bicycle	plant

Reproducible page from *Exploring Energy with **TOYS*** published by Terrific Science Press™

The Energy Transformation Game

Mystery Cards for Full-Size Game

R,E → S,L television	C → M,H person	L → C plant
E → S siren	C → S,L firecracker	M → M bicycle
R → E satellite dish	E,M → S electric guitar	C → L lightstick
C → M backhoe	C → H,L candle	E → H iron
L → E solar calculator	L → H magnifying glass	E → M mixer
E → R,H microwave oven	H → M hot-air balloon	E → L,H light bulb
M → E windmill	N → E,H nuclear power plant	M → S piano
N → H,L,R sun	E → R tanning bed	C → E battery 9 Volt

Mystery Cards for Travel-Size Game

N → H,L,R sun	M → E windmill	E → R,H microwave oven	L → E solar calculator	C → M backhoe	R → E satellite dish	E → S siren	R,E → S,L television
E → R tanning bed	N → E,H nuclear power plant	H → M hot-air balloon	L → H magnifying glass	C → H,L candle	E,M → S electric guitar	C → S,L firecracker	C → M,H person
C → E battery	M → S piano	E → L,H light bulb	E → M mixer	E → H iron	C → L lightstick	M → M bicycle	L → C plant

The Energy Transformation Game

Clue Check-Off Sheets

Energy Guess Who Clues	
Converts FROM	Converts TO
electrical	electrical
sound	sound
radiant	radiant
mechanical	mechanical
heat	heat
light	light
nuclear	nuclear
chemical	chemical

Energy Guess Who Clues	
Converts FROM	Converts TO
electrical	electrical
sound	sound
radiant	radiant
mechanical	mechanical
heat	heat
light	light
nuclear	nuclear
chemical	chemical

Energy Guess Who Clues	
Converts FROM	Converts TO
electrical	electrical
sound	sound
radiant	radiant
mechanical	mechanical
heat	heat
light	light
nuclear	nuclear
chemical	chemical

Energy Guess Who Clues	
Converts FROM	Converts TO
electrical	electrical
sound	sound
radiant	radiant
mechanical	mechanical
heat	heat
light	light
nuclear	nuclear
chemical	chemical

The Energy Transformation Game

Blank Cards for Full-Size Game

Blank Cards for Travel-Size Game

Drop 'n' Popper

...Students make their own Poppers and investigate the effect of surface type on Popper jump height.

Commercial and homemade Poppers

✔ Time Required

Setup   10 minutes
Performance   45 minutes
Cleanup   5 minutes

✔ Key Science Topics

- elastic potential energy
- force
- gravitational potential energy
- kinetic energy
- work

✔ Student Background

Students should be familiar with the transformation of gravitational potential energy to kinetic energy, as explored in the activity "Bounceability." Students should also have experience making observations, using a meterstick to measure height, and recording data.

✔ National Science Education Standards

Science as Inquiry Standards:

- Abilities Necessary to Do Scientific Inquiry

  Students execute an investigation of the effect of surface texture on Popper jump height.

  Students use their results to explain how the surface on which the Popper is dropped affects energy transformations in the Popper.

Physical Science Standards:

- Motions and Forces

  Unbalanced forces cause changes in an object's velocity.

- Transfer of Energy

  Energy is associated with mechanical motion, deformation of objects, and sound. Work done by forces is a means of transferring energy.

✔ Additional Process Skill

- measuring   Students measure the Popper's jump height.

MATERIALS

For Introducing the Activity
- (optional) racquetball
- (optional) several potential-to-kinetic energy toys, such as spring-loaded pop-ups
- (optional) Popper made according to Part A of the Procedure or commercially available popper

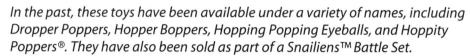

In the past, these toys have been available under a variety of names, including Dropper Poppers, Hopper Boppers, Hopping Popping Eyeballs, and Hoppity Poppers®. They have also been sold as part of a Snailiens™ Battle Set.

For the Procedure
Part A, per group of 2–4 students
- racquetball

One ball will make two poppers.

- sharp knife
- heavy-duty, sharp scissors

Utility scissors with serrated blades work very well for this purpose. If you use regular heavy-duty scissors, you may need to trim in very thin layers.

- 3-inch x 5-inch index card
- marker
- (optional) tape

Part B, per group of 2–4 students
- Popper made in Part A
- test surface squares approximately 30 cm x 30 cm, of the following materials
    - short-nap carpet
    - long-nap carpet
    - flooring tile
    - other materials, such as ceiling tile, cardboard, cork, foam, wood
- 2 metersticks
- (optional) masking tape

For the Extensions
❷ All materials used for Part A of the Procedure, except
Per group of 2–4 students
- substitute a tennis ball for the racquetball

❹ Per class
- Grabbin'™ Grasshoppers game by Tyco®

SAFETY AND DISPOSAL

Remind students to use caution when cutting and trimming the racquetballs. No special disposal procedures are required.

GETTING READY

Make several Poppers prior to class as examples of the finished product. If students are not going to make their own Poppers, make enough for the class according to the directions in Part A of the Procedure. Students in seventh grade and up should be able to make their own Poppers, although class time could be saved by having a few volunteers prepare them ahead of time. Most fifth and sixth graders will not be able to make the Poppers without significant assistance.

INTRODUCING THE ACTIVITY
Options:

- If your students have completed the activity "Bounceability," review the results in terms of energy conversions.

- Drop a racquetball and observe that the ball does not bounce back to the height from which it was dropped. Drop a Popper and observe that it jumps back to a height greater than that from which it was dropped. Discuss the transformations from potential energy to kinetic energy in each example. Discuss other variables that might affect Popper jump height.

- Demonstrate other examples of potential-to-kinetic energy toys, such as spring-loaded pop-ups, and discuss the energy conversions.

PROCEDURE
Part A: Making the Popper

Have students make Poppers:

1. Trace a circle with a diameter of 5½ cm in the center of a 3-inch x 5-inch index card. A template is provided in Figure 1.

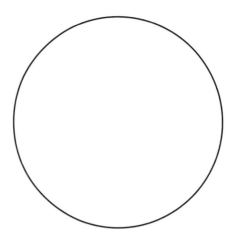

Figure 1: Template for the Popper

2. Carefully cut out the circle on the card. Discard the circle and keep the remainder of the card to use as a template for cutting the Popper. If you started cutting from the edge of the card instead of from the inside of the circle, tape the edges of the beginning cut together to help the template hold its shape.

3. Locate the racquetball seam. Using the knife, cut the racquetball in half along the seam line.

4. Place one racquetball half on the desktop with the cut edge facing downward. Place the template like a hat on the racquetball with the template parallel to the seam of the ball. Push the template down until no space is left between the ball and the template. (See Figure 2.) Make sure the template is straight on the ball (an even distance from the edge all around). If necessary, turn the ball and card upside down to check and straighten the template.

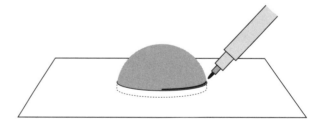

Figure 2: Push the template over the ball until there is no space between the ball and the card. Draw a line to mark where the ball and card meet.

5. Using the marker, mark a ring around the ball where the template and the ball meet. (See Figure 2.) Remove the template. The ring should be approximately 0.5–1.0 cm from the cut edge.

6. Invert the ball by pushing inward on the top with the thumbs. Using the scissors, cut along the outside of the marked line, that is, on the side of the line toward the cut edge. Discard the trimmings.

7. Keeping the Popper inverted, drop it with the now-open side downward onto a hard-surfaced floor, such as a tile or hardwood floor. The Popper should jump back much higher than its initial drop height. (See Figure 3.)

 Trimming your Popper to its most effective size requires quite a bit of trial and error. If you drop your Popper and it doesn't pop up higher than its drop height, follow the instructions in step 7a or 7b. Making the first successful Popper may take a while, but subsequent Poppers will take much less time.

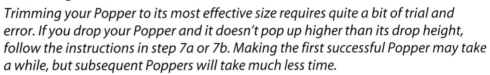

Figure 3: The Popper on the left is ready to be inverted. The Popper on the right is inverted and ready to jump.

a. If the Popper does not jump into the air upon contacting the floor, try dropping it from a higher distance. If it still doesn't pop, CAREFULLY trim a SMALL amount of rubber from the cut edge, using the marked line as a guide.

 Make sure the lip of the Popper forms an even circle. If one section bulges out noticeably, try trimming just that section.

b. If the Popper flips back to its original shape from its inverted configuration before it reaches the floor, too much rubber was trimmed off the Popper, perhaps because the template was not down far enough on the ball when the line was marked. In this case, repeat steps 4–7 with a new racquetball half. However, you may want to save the overtrimmed Popper; it may work for Extension 2.

Part B: The Effect of Surface Type on Popper Jump Height

Have each group investigate the effect of surface type on jump height according to the following steps. Each group should test all available surface types.

Each group will need a Dropper, Meterstick Holder/Surface Changer, Estimator, and Recorder to perform the following jobs:
* *Dropper—to invert and drop the Popper;*
* *Meterstick Holder/Surface Changer—to hold the metersticks and change the surface being tested;*
* *Estimator—to eyeball the height to which the Popper jumps; and*
* *Recorder—to record the jump heights on the Data Sheet (provided).*

1. Make a 2-meter stick by placing the zero end of one meterstick on the floor and placing the zero end of another meterstick on top of the first meterstick. (The Meterstick Holder should hold the metersticks in place throughout the investigation.) Masking tape may be used to attach the two metersticks if desired.

2. Invert the Popper and drop it from a height of 1 m. When the Popper pops, eyeball the height to which it rises. Repeat several times to develop consistency in estimating the height. (A somewhat higher initial height may be used if necessary to get the Popper to pop consistently.)

3. Complete three trials, record the jump height for each in centimeters on the Data Sheet, and calculate and record the average height.

4. Repeat steps 1–3 for each test surface.

5. Answer the questions on the Data Sheet.

6. Discuss the questions on the Data Sheet as a class. Develop a class explanation of how the Popper works.

EXTENSIONS

1. Vary the height from which the Popper is dropped instead of varying the surface on which the Popper is dropped. Notice that Popper jump height does not vary as much as you would expect with initial dropping height. This indicates that the stored elastic potential energy is more important than the initial gravitational potential energy in determining jump height. The fact that the Popper bounces very poorly if dropped without being inverted also confirms this conclusion.

2. Instead of dropping the Poppers, set inverted Poppers open side downward on various surfaces and watch them pop up. How do these pop heights compare with the differences between drop height and jump height in Part B of the Procedure? (You may also want to try overtrimmed Poppers from step 7b of the Procedure.)

3. Make a Popper from a tennis ball using the same procedure as for the racquetball but using a template 5.9 cm in diameter. However, trim with extreme caution since the tennis-ball Poppers flip back to their original shape more easily than those made from racquetballs. (See Part A of the Procedure, step 7.) The tennis-ball Popper does not pop as high because of the fuzz on the ball. (See the Explanation.)

4. Play Grabbin' Grasshoppers, by Tyco. Explain the physics of the jumping grasshoppers. Why do the instructions suggest rubbing the playing board with waxed paper?

EXPLANATION

The following explanation is intended for the teacher's information. Modify the explanation for students as required.

The action of the Popper can be explained in terms of energy transformations. The energy transformations of the Popper involve two actions: (1) the Popper acting as a bouncing rubber ball and (2) the popping action of the Popper. The bouncing and popping actions take place at the same time when the Popper contacts the floor.

Let us first consider the energy transformations in a dropped ball. Before it is dropped, the ball has a certain amount of gravitational potential energy that is proportional to its height above the floor. (Gravitational potential energy = mgH, where m is the mass of the ball, g is the acceleration produced by gravity (9.8 m/s^2), and H is the height above the floor.) As the ball falls, its gravitational potential energy is converted into kinetic energy. (Kinetic energy = ½ mv^2, where v is velocity.) When the ball makes contact with the floor surface, this kinetic energy is converted into sound energy, thermal energy (which affects both ball and surface), and elastic potential energy (deformation energy) as both the ball and the surface are deformed. Some of this deformation energy, but not the sound and heat energy, is converted back to kinetic energy as the ball leaves the surface. How much of the deformation energy is regained as kinetic energy is determined by how fast the surface and ball return to their original shapes. If portions of the surface or ball remain deformed after the ball leaves the surface, that part of the deformation energy does not contribute to the ball's kinetic energy. This kinetic energy is eventually converted into gravitational potential energy as the ball rises and reaches its maximum height. This maximum height resulting from the bouncing action alone is less than the original height, as not all of the ball's original gravitational potential energy is transformed back into its final gravitational potential energy.

Now let us analyze the popping action of the Popper. This action takes place at the same time as the bouncing action described above. The process begins when the Popper is inverted by pushing inward on the rounded dome with the thumbs. Work is done on the Popper as force is used to depress the dome of the Popper through a distance. (Work = force times distance.) This work gives the Popper elastic potential energy, or deformation energy. When the Popper hits the floor, this stored potential energy is converted to kinetic energy as the inverted dome of the Popper moves toward the floor.

When the dome of the Popper contacts the floor, it exerts a downward force on the floor. According to Newton's third law, the floor exerts an equal and opposite upward force on the Popper. The force of the floor on the Popper results in work being done on the Popper by the floor. This work causes the Popper to acquire kinetic energy and an upward velocity.

The Popper can achieve a greater height than a ball since it has greater kinetic energy as it leaves the floor as a result of contributions from both bouncing and popping actions. The maximum height reached by the Popper can be greater than the initial dropping height because of the additional energy gained from the initial work done on the Popper when the dome is inverted. As discussed in the energy analysis of the Popper as a bouncing ball, some of the stored elastic potential energy may not be recovered in the form of kinetic energy if portions of the ball or drop surface do not return to their original shapes before the ball leaves the surface. This results in different final heights for different surfaces and for Poppers made of different kinds of balls—for example, lower heights would result from the slow responses of the compressed nap of a carpet and of fuzz on a tennis ball.

ASSESSMENT

Options:

- Use students' completed Data Sheets to assess student understanding.

- Have students use what they have learned to explore the action of a tennis-ball Popper:
  - Examine and drop a tennis ball.
  - Write a hypothesis for what will happen when you drop a tennis-ball Popper.
  - Make a tennis-ball Popper as described in Extension 3. Test the tennis-ball Popper on a tile or hardwood floor.
  - Compare the average maximum heights of the racquetball and tennis-ball Poppers and explain any differences in the average maximum heights.

CROSS-CURRICULAR INTEGRATION

Language arts:
- Have students invent, describe, and write instructions for a new game using the Popper.

Math:
- The activity involves measuring and making a data table, and it could also involve creating a bar graph. Have students determine and discuss the ratio of Popper jump height to initial drop height.

REFERENCE

Guzdziol, E.S. "Astounding Jumping Disk," *Science Scope*. 1991, 15(3), 28–30.

CONTRIBUTOR

Lee Ann Ellsworth, Northwestern Middle School, Springfield, OH; Teaching Science with TOYS, 1991–92.

HANDOUT MASTER

A master for the following handout is provided:
- Data Sheet

Copy as needed for classroom use.

Names _____   _____

_____   _____

Drop 'n' Popper
Data Sheet

	Data Table: Popper Jump Height (cm)				
	Surface				
	Tile	Short-nap Carpet	Long-nap Carpet	_____	_____
Trial 1					
Trial 2					
Trial 3					
Sum					
Average					

Questions

Answer the following questions on a separate sheet of paper.

1. Work is done when a force is exerted through a distance. When do you do work on the Popper?

2. What kind of energy does the Popper receive from the work done in inverting it?

3. When does the Popper have gravitational potential energy?

4. Summarize how energy is involved in the way the Popper works. Be sure to explain how this is both similar to and different from the way an ordinary racquetball bounces.

5. Does the type of surface affect the conversion of potential energy to kinetic energy? Discuss your answer using the information from your data table. What characteristic of the surface seems to be the most important variable?

    Reproducible page from *Exploring Energy with TOYS* published by Terrific Science Press™

Apply Your Energy Knowledge

...Students observe a variety of ways in which energy can be stored and released and apply knowledge about kinetic and potential energy to stored-energy toys.

✔ **Time Required**

Setup   15–45 minutes*
Performance 60–75 minutes (may be spread
                       over several days)
Cleanup   10 minutes

 *The long end of this range includes time for preparing the answer key for the Learning Center Method.

Energy-storing toys

✔ **Key Science Topics**

- elastic potential energy
- forces
- gravitational potential energy
- kinetic energy
- work

✔ **Student Background**

Students should be very familiar with the concepts of work, forces, energy, kinetic energy, and potential or stored energy. The activity will give the students an opportunity to review, apply, and further explore these concepts.

✔ **National Science Education Standards**

Science as Inquiry Standards:

- Abilities Necessary to Do Scientific Inquiry

  Students observe the motion of a toy and describe its motion in terms of the energy transformations taking place.

  Students communicate their descriptions to the rest of the class.

Physical Science Standards:

- Transfer of Energy

  Energy is associated with mechanical motion and is transferred in many ways.

MATERIALS

Per class:

- toys that store and release energy—some possibilities are

Spring-ups	Flip Frog
Push-n-Go®	Popovers®
See-Thru-Loco	Bath Tubbie
Turblo	dart gun
Explorer Gun®	pull-back car
High Flyer	any battery-operated toy
Ping-Pong™ gun	

SAFETY AND DISPOSAL

No special safety or disposal procedures are required.

GETTING READY

If using the Learning Center method, fill out a sample Data Sheet (provided) for each toy. This completed Data Sheet will be the Sample Answer Key.

INTRODUCING THE ACTIVITY

This activity can be done in small groups all working at the same time or by individuals or pairs working in a Learning Center over several days. In either case, you should introduce the activity by going through the steps of the Data Sheet for one toy with the class as a whole to make sure everyone understands the questions. You might want to make transparencies of the Data Sheet to use during this discussion.

PROCEDURE

Small Group Method:

1. Divide students into groups of four and assign jobs.

   Each group will need a Writer, a Go-fer, a Reporter, and an Organizer/Encourager to perform the following jobs:
   - *Writer—to record the group's answers on the Data Sheet to be turned in,*
   - *Go-fer—to get materials from the teacher and also to interact with the teacher if the group has a question,*
   - *Reporter—to represent the group in the class discussion and to help the Writer phrase the answers for the Data Sheet, and*
   - *Organizer/Encourager—to keep the group on task and to encourage everyone to participate. Students may need to be trained in how to do this by specifically asking for an opinion from someone who hasn't said much or praising contributions with phrases like "That's right" or "Good idea."*

2. Ask Go-fers to collect a Data Sheet and a toy for their group.

3. Have the groups experiment with their toys and complete the Data Sheet. You can circulate and answer questions. Some groups will get so interested in investigating what their toy will do (for example: Will it climb a hill? Will it jump the gap between two desks?) that they will forget to fill out the Data Sheets. You may want to tell them that these investigations are permitted but only after the Data Sheet is completed and only if carried out as a controlled scientific experiment, not just as play.

4. After all groups have finished, have a Reporter from each group demonstrate the group's toy for the class and describe the energy transformations that take place. After all the groups have presented, ask the class what generalizations might be made about all the toys. Students should be able to come up with ideas such as, "Some kind of energy must be input in order for the toy to move," and "By the end, kinetic energy has turned into thermal energy or some other kind of energy, such as sound."

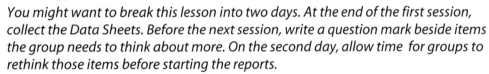
You might want to break this lesson into two days. At the end of the first session, collect the Data Sheets. Before the next session, write a question mark beside items the group needs to think about more. On the second day, allow time for groups to rethink those items before starting the reports.

Learning Center Method:

1. Give students the following instructions for using the Learning Center:

   a. Choose a toy from the center, get a Data Sheet, and investigate the toy.

   b. Record observations and answer the questions on the Data Sheet.

   c. Turn in the Data Sheet. After the Data Sheet is returned, check the Sample Answer Key (see Getting Ready) and compare with your Data Sheet. If desired, add additional comments to the Data Sheet in a different color of ink or pencil than was used originally.

   d. Repeat steps a–c with at least two other toys.

2. After the center explorations are complete, hold a class discussion to help students summarize their observations. Briefly demonstrate and discuss each toy so that all students may see all of the toys. You may want to call on someone who tested each toy to discuss it for the class. Then, focus the discussion on similarities and differences between the energy toys and the generalizations that may be made about these observations.

Since this discussion is presented as a review of concepts, one class period or less should suffice.

EXPLANATION

The following explanation is intended for the teacher's information. Modify the explanation for students as required. You may want to read the Data Sheet before reading this explanation.

Obviously, a specific explanation cannot be provided for every toy that could possibly be included in this activity. However, some explanations that can easily be adapted to a variety of toys are provided below.

Explorer Gun

The kinetic energy (KE) you observe with this toy is in the motion of the flying disk. This energy is supplied by your muscles turning the disk. This energy is stored as elastic potential energy (EPE) in a spring in the toy. It stays there until you pull the trigger, which allows the spring to unwind. The energy stored in the spring is converted to the KE you observe in the disk's flight. You can change the amount of KE from one trial to another by winding the disk more or fewer times. If you give the toy more energy, more energy is transferred to the disk. The toy has the most EPE when you have wound the disk as far as you are going to wind it. When the disk has the most gravitational potential energy (GPE) depends on whether you fire it horizontally or at an angle. If fired horizontally, the disk has the most GPE at the very beginning of the flight. If fired at an angle, it has the most GPE when it is at its highest point. Because of the action of gravity, the issue of when KE is greatest is complicated. In general, if students are able to back up their answers with reasonable explanations, you may accept either of two answers: when the disk leaves the gun or just before it hits the floor.

Spring-Ups

Before the toy begins moving, energy is stored as elastic potential energy in the toy's spring when the energy of your muscles pushes down on the toy and makes the suction cup stick to the base. The toy has the most EPE when the spring is fully compressed. When the suction cup releases, the EPE is converted to kinetic energy. The toy has the most KE when the spring is completely expanded. You cannot change the height of the jump or the toy's speed from one trial to the next; the amount of initial KE is constant. As it moves up, the toy slows because gravity is pulling it down. As this happens, the KE is being changed into gravitational potential energy.

Spring-ups often have a flipping motion because the spring often bends slightly as the suction cup releases. As a result, the force exerted is not perfectly vertical. At its highest point, almost all of the toy's kinetic energy is converted into gravitational potential energy. In most cases, the toy will still have some rotational motion, so it must still have some KE. As the toy comes back down, the GPE is converted back to KE. When the toy hits the table and stops, it loses both its potential and kinetic energy. Where does the energy go? Primarily, it becomes thermal energy, but some of it goes into sound energy.

Flip Frog/Popover

The frog moves, so it must have kinetic energy. This energy comes from your muscles when you press down on the frog's head. This muscle energy is then stored in the spring in the frog as elastic potential energy. The energy is stored there until the suction cup releases. This causes the EPE in the spring to be converted to KE, the energy of the frog's jump. You cannot adjust the amount of initial KE from one trial to the next. The KE is determined by the spring and the suction cup. This toy has the most KE when it first begins to jump. It has the most EPE when you finish pushing it down and attaching it to the base (before it jumps).

The motion of the Popover is similar. Instead of stretching a spring, you put energy in by winding a spring. When you release the winding stem, the spring begins to unwind, stretching another spring just as in the frog. Instead of a suction cup release to initiate the flip, a gear mechanism inside the Popover releases when the spring has been stretched a certain amount.

Both toys' motion is in two parts: up and down, and backwards in a vertical circle (the flip). The up-and-down motion can be explained just like the Spring-Up's motion. The rotational motion is a little harder to explain. When the spring is released, it moves inward at both ends, pulling the head backward and the legs upward. Since the legs are off-center, this causes the toes to push down on the table, and in an example of action-reaction forces, the table in turn exerts an upward force on the toes. This causes the gorilla or frog to flip backwards, turning around its center of mass.

High Flyer

This toy is excellent for illustrating the scientific definition of work (force times distance). As you pull the string, you can clearly feel the reaction force of the string pulling back on your hand. Your pulling force, exerted as the string moves, does work that is then transferred to energy of the Flyer. Most of this energy is given directly to the flying disk, but some is stored in a twisted rubber band in the handle. When you release the cord, this stored energy is used to pull the cord back into the handle. The greater the distance you pull the string before releasing it, the greater the energy of the disk. The moving disk has both kinetic energy and gravitational potential energy. It has the most GPE when it is at the highest point in its flight. It has the most KE when its speed is greatest, but it is difficult to pinpoint exactly when that is. It might be as the disk leaves the launcher, or it might be just before it hits the ground.

Push-n-Go®

When you push the rider's head down, this toy moves forward. This is an example of kinetic energy. Another example of KE is the upward movement of the rider's head after you remove your finger. The energy originally comes from you as you push down on the rider's head. You exert a force while moving the head, so work is done transferring energy from you to the toy. The energy is stored as elastic potential energy for a moment in the spring inside the toy. When you let go, the spring extends, and its PE is converted to KE. The toy has the most KE immediately after the head returns to its original position. The toy has the most EPE when you have pushed the head all the way down and haven't yet released it. Since the toy never leaves the ground, it has no gravitational potential energy (unless you send it up a hill). Eventually the toy does stop moving, so it no longer has any KE. The force of friction between the wheels and floor causes the toy to stop. As part of this process, the toy's KE is converted to thermal energy. (Further explanation related to this toy can be found in the activity "What Makes It Go?")

See-Thru-Loco

Energy is supplied in this toy by your muscles when you wind the spring. This energy is stored in the spring as elastic potential energy and is stored there until you release the key. Then the potential energy is converted to the kinetic energy of the toy's movement. The entire locomotive moves, and its internal parts also move. Both of these movements involve KE. This toy has the most PE when you have finished winding the key and haven't yet released it. The toy has the most KE when it is moving fastest— somewhere in the middle of the motion. The force of friction between the tires and the floor and between the internal parts causes the toy to slow down and eventually stop. The toy's KE is turned into thermal energy—the toy and the floor both get a little warmer.

The gears transfer the energy from the spring to the wheels of the locomotive. The function of each gear should be identified. Notice that different gears turn at different speeds. To see the gears turn in slow motion, wind up the toy and then slowly turn the winding key backwards. (Further explanation related to the operation of gears can be found in the activity "Get It In Gear with a Lego® Vehicle.")

Battery-Operated Toys

Although many different types of battery-operated toys are available, some general statements can be made about all of them. Batteries are storage devices for chemical energy. When the battery is placed in a complete circuit, the chemical energy is gradually converted to electrical energy. This electrical energy may be used to cause motion or to produce light, sound, or a variety of other forms of energy.

VARIATION

Divide the students into small groups and assign each group a toy to demonstrate and explain. Then have the class evaluate each group's presentation by answering the following questions:

1. Did the group explain all of the energy forms used? Explain.

2. Were you able to understand their explanations?

3. On a scale of 1 to 5, with 5 being the highest rank, how would you rank this presentation? Make comments to explain your ranking.

ASSESSMENT

Options:

- The group Data Sheets may be collected and evaluated after the presentations. The presentations themselves could also be evaluated for correctness and completeness.

- At a later time, students could be asked to analyze a new toy using the Data Sheet, which would then be graded. This could be done as an individual or group assessment and is particularly appropriate as an end-of-unit assessment.

CROSS-CURRICULAR INTEGRATION

Earth science:

- After observing in detail how toys use and convert energy, students may be much more perceptive in recognizing energy use, energy transfer, and energy waste by humans. As either a follow-up to this lesson or an introduction to the next area of study, present an environmental science lesson regarding the ways and reasons why humans use energy. To introduce the lesson, have the students keep a brief energy diary, recording ways in which they have used energy from within their own bodies and from other sources during one 2-hour period.

Language arts:

- The process of comparing energy toys lends itself to writing exercises involving comparison and contrast.
- Have students write stories about what would happen if one of the energy toys that resembles an animal or person came to life or could talk.
- Students could write a persuasive piece—perhaps a radio or TV commercial—encouraging fellow students to use toys whose energy comes from the people playing with them, rather than from batteries.

Social studies:
- Have students study life in an earlier time, such as the pioneer era, ancient Rome, or the Middle Ages, and compare typical toys from that time to typical toys today. Encourage them to note how these historical toys stored and released energy.

FURTHER READING

Faughn, J.; Turk, J.; Turk, A. *Physical Science;* Saunders: Philadelphia, PA, 1991. (Teachers)

Gartrell, J.E.; Schafer, L.E. *Evidence of Energy;* National Science Teachers Association: Washington, D.C., 1990. (Teachers)

Kirkpatrick, L.D.; Wheeler, G.F. *Physics: A World View,* 2nd ed.; Saunders: Philadelphia, PA, 1995. (Teachers)

CONTRIBUTOR

Anita Kroger, Gifted and Talented Specialist, Cincinnati, OH; Teaching Science with TOYS, 1988–89.

HANDOUT MASTER

A master for the following handout is provided:
- Data Sheet

Copy as needed for classroom use.

Name _____   Date _____

Toy used _____

Apply Your Energy Knowledge
Data Sheet

Use the toy several times and observe carefully before answering these questions.

1. Describe in your own words what this toy does.

    _____

    _____

    _____

    _____

2. Does the toy's speed change? _____   Does its direction change? _____
    If you answered yes to either or both of these questions, explain what caused the change(s).

    _____

    _____

    _____

3. When does this toy have kinetic energy?

    _____

    _____

4. Where did this energy come from? (For example, it may come from batteries that store chemical energy.)

    _____

    _____

    _____

5. An object has the most kinetic energy when it is moving the fastest. At what point in its motion does this toy have the most kinetic energy?

    _____

6. Can you make the amount of initial kinetic energy change from one trial to the next? If so, how?

_____

_____

7. Was gravitational potential energy or elastic potential energy ever stored in this toy? If so, where or how was it stored?

_____

_____

_____

If your answer to #7 was No, skip questions 8, 9, and 10.

8. What caused the stored energy to be converted to kinetic energy?

_____

_____

9. When does the toy have the most elastic potential energy? (Skip this question if it does not have any.)

_____

_____

10. When does this toy have the most gravitational potential energy? (Skip this question if it does not have any.)

_____

_____

11. Use the space below to record any observations you made about this toy that are not included in your answers to the Data Sheet questions.

_____

_____

_____

_____

Doc Shock

...Use the Operation game to explore the energy transfers in an electric circuit.

✔ Time Required

Setup	10	minutes
Performance	45	minutes
Cleanup	5	minutes

Operation game

✔ Key Science Topics

- conductors and nonconductors (insulators)
- electricity
- energy transfer
- open and closed circuits
- parts of a circuit

✔ Student Background

Students should have been introduced to the concept of the electrical circuit via some experimentation with batteries and bulbs. "Doc Shock" is then appropriate as an application activity on electric circuits or energy transformations.

✔ National Science Education Standards

Science as Inquiry Standards:

- Abilities Necessary to Do Scientific Inquiry

  Students form hypotheses about why the game board lights up and buzzes as they play.

  Students observe and investigate the internal parts of the game.

  Based on their observations, students explain how the game works.

Physical Science Standards:

- Transfer of Energy

  Electrical circuits transfer energy from the battery to the light bulb and buzzer, where it is converted to light and sound energy.

MATERIALS

For Getting Ready only
- knife

For the Procedure
Per group of 4 students
- 2 D-cell batteries
- Operation® game by the Milton Bradley Company

> *This game is available at toy stores, and some students may have this game at home.*

For the Extensions
❶ All materials listed for the Procedure plus the following:
Per class
- assorted small objects to test for conductivity, such as pencils, paper clips, pennies, erasers, and plastic construction blocks

> *The objects must be able to be held by the game's tweezers.*

❷ All materials listed for the Procedure plus the following:
Per class
- piece of wire about 10 cm long (A straightened small paper clip works well.)

❸ Per class
- other games or toys in which bulbs light or buzzers buzz

❹ Per group
- paper clips
- cardboard
- insulated wires
- aluminum foil
- D-cell batteries
- flashlight bulbs

❺ Per group
- buzzer or bulb

> *The doorbell described in* Dear Mr. Henshaw *may be hard to find, but Radio Shack sells a variety of buzzers that could be used.*

- battery

> *Check the packaging of the bulb or buzzer for the voltage of battery required.*

- wire
- switch (commercial or homemade)
- (optional) container with lid
- (optional) tape

SAFETY AND DISPOSAL

No special safety or disposal procedures are required.

GETTING READY

The cardboard cover of the Operation game board is held down by plastic pegs. Use a knife to cut the cardboard around the pegs. If students are bringing the game from home, check with parents for permission to remove the cover. The cover can be replaced easily, restoring the game to full usefulness.

INTRODUCING THE ACTIVITY

Explain the game and its rules or allow students who have played it before to explain the game to the class.

PROCEDURE

1. Explain to the students that they will be playing the Operation game in small groups. Ask them to think about why the game board is making noise and lighting up as they play.

2. After they play, have the students stop and share with a partner their thoughts about the reasons behind the noise and light or have them write their ideas down.

   Do not hold a class discussion until step 5.

3. Have each group of students take the cover off their game board and investigate the internal parts. Ask groups to make the bulb light and the buzzer buzz while observing all the internal parts involved in these actions. Have each group list these parts, trace the path the electric current follows, and draw a picture showing the path. If students have previously studied series and parallel circuits, they should determine which type of circuit is used in this game.

4. Have the students share their observations about the necessary components of the game in a class discussion and list the common observations on the board. Match these with general terms that apply to any circuit, as shown below.

| Table 1: Classifications of the Parts of an Operation Game ||
General Components of a Circuit	Specific Components of the Operation Game
energy source	battery
energy paths	tweezers metal pieces of board (not plastic or cardboard) wires
energy receivers	bulb and buzzer

5. Ask a few students to describe their initial hypotheses about how the game works and discuss whether their investigation supported or refuted their hypotheses.

6. Summarize the activity as follows: "All of the components on our list are necessary for an electric circuit to work. If they are all connected and working, the circuit is closed, or completed, and electric current moves around the entire path. If any connections are not made, the circuit is open, and electric current can't move around the entire path. An open circuit is like a raised drawbridge—nothing can move across. When the circuit is closed, energy can be transferred from the battery to the light bulb and buzzer."

EXPLANATION

The following explanation is intended for the teacher's information. Modify the explanation for students as required.

Every electric circuit has three essential components: a source of electrical energy, a receiver or user of electrical energy, and one or more objects through which the electric current may travel to get from the source to the receiver and back again. The most easily recognized energy source is the battery. The source could also be a generator, or it could be an outlet in the wall through which electricity comes from a generator that might be far away. The receiver could be, for example, a bell, buzzer, light bulb, toaster, curling iron, or alarm clock. The most common object through which electricity travels in an electric circuit is a covered copper wire, but it could be a metal plate, water, or a human being. An object that works in this capacity—one through which electricity will travel—is called a conductor. An object through which electricity will not travel is called a nonconductor, or insulator. A circuit may contain additional components, such as a switch. A switch is simply a handy device for opening or closing the circuit without disturbing any of the other connections.

The Operation game uses a simple electric circuit. When the cover is removed from the game, the circuit's basic components can be seen. The two batteries act as the energy source; the metal plate and the two wires are the path through which the energy flows. The bulb and buzzer are the energy receivers, and the tweezers act as a switch, completing the circuit when they touch the metal plate. Figure 1 shows a diagram of this circuit.

Exploring Energy with **TOYS**

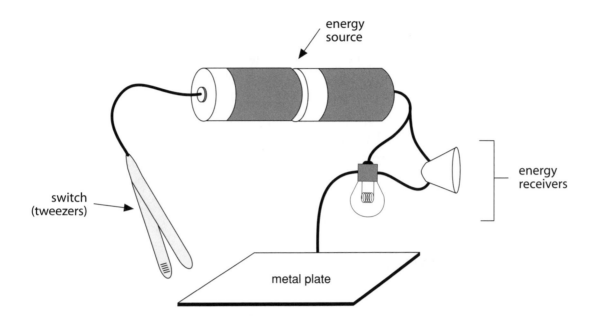

Figure 1: The Operation game uses a simple electric circuit.

A schematic diagram of this circuit would resemble the one shown in Figure 2.

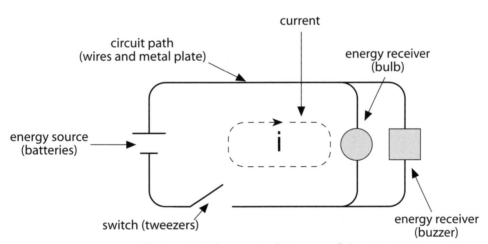

Figure 2: Schematic diagram of the circuit

When the switch is closed (either side of the tweezers touches the metal plate), the circuit path is closed and the current (i) flows. This is a parallel electric circuit because current can get to the bulb without going through the buzzer and vice versa. In contrast, in a series circuit, all the current must go through each element. An example of a series circuit is a string of old-fashioned Christmas lights; when one bulb burns out, the circuit is broken, and no bulbs in the strand will light.

Energy transformations occur in both the bulb and the buzzer. The filament of the bulb is made of a type of wire that gets very hot when electric current passes through it. Thus, electric energy is converted to heat, or thermal energy. This energy is quite obvious if you touch a light bulb that has been on for a while. When the filament gets hot enough it begins to glow, emitting visible light. Light is itself a form of energy, so energy is transferred from the bulb to you as the light enters your eye. In the buzzer part of the circuit, the electric current operates an electric motor, causing a shaft to spin. The shaft hits a metal plate, causing it to vibrate. Thus, electrical energy has been converted to kinetic energy of both the shaft and the metal plate. As it vibrates, the plate pushes against nearby air molecules, producing sound waves. It has given some of its kinetic energy to the air molecules. The sound waves carry this energy to your ear.

EXTENSIONS

1. While the game board is disassembled, allow students to test other objects to see whether they will transfer electrical energy. To do this, hold objects with the tweezers and attempt to make a closed circuit by touching the objects to the metal plate. Introduce the terms "conductor" and "insulator."

2. Remove one of the batteries from the game and use a wire or straightened paper clip to complete the circuit in place of the removed battery. Ask the students to explain why the bulb is dimmer and the buzzer slower.

3. Have the students bring in other games or toys in which bulbs light or buzzers buzz in order to see how they are constructed.

4. Have students make their own game boards with paper clips, cardboard, insulated wires, aluminum foil, batteries, and bulbs.

5. Have the students read *Dear Mr. Henshaw,* by Beverly Cleary (Avon, ISBN 0380709589). If time doesn't permit reading the entire book, focus on the diary entries for Thursday, March 1, through Monday, March 5. A circuit similar to the one Leigh builds can be constructed as either a group or class activity. Have the students demonstrate that they understand how and why switches and electric circuits work. After the group project is completed, leave the materials out in a learning center with reminder cards so that students may individually experiment and make their own alarms.

CROSS-CURRICULAR INTEGRATION

Home, safety, and career:
- Have students write a report or develop public service materials about electrical safety in the home. Topics may include how humans can be conductors of electricity, making outlets safe from small children, not overloading circuits, keeping electrical appliances away from water, etc.

Language arts:
- After completing the activity, have the students write a simple descriptive paragraph about an electric circuit, describing how it works, what parts it includes, and what it can be used to do. Later you may wish to have the students write a comparison/contrast paragraph between open and closed circuits or other similar, but not identical, circuits.

FURTHER READING

Ardley, N. *Discovering Electricity;* Franklin Watts: New York, NY, 1984. (Students)

Faughn, J.; Turk, J.; Turk, A. *Physical Science;* Saunders: Philadelphia, PA, 1991. (Teachers)

Kirkpatrick, L.D.; Wheeler, G.F. *Physics: A World View*, 2nd ed.; Saunders: Philadelphia, PA, 1995. (Teachers)

Thier, H.D.; Knolt, R.C. *SCIS III: Scientific Theories;* Delta Education: Hudson, NY, 1993. (Teachers)

CONTRIBUTORS

Holly Rice, Wyoming Middle School, Wyoming, OH; Teaching Science with TOYS, 1988–89.

Hope Spangler, Graduate Student, Physics Department, Miami University, Oxford, OH, 1991–93.

Make Your Own Motor

...Students use wire, a battery, and a magnet to make a simple electric motor.

✔ **Time Required**

Setup	10	minutes
Performance	30	minutes
Cleanup	5	minutes

✔ **Key Science Topics**

- electric current
- electric motor
- energy transformations
- magnetism

✔ **Student Background**

Students should have been introduced to the concepts of permanent magnets, batteries, electric current, and kinetic energy.

A mini-motor

✔ **National Science Education Standards**

Science as Inquiry Standards:

- Abilities Necessary to Do Scientific Inquiry

  Students identify variables which may affect the performance of the motor.

  Students investigate the effects of some of these variables in controlled experiments.

Physical Science Standards:

- Transfer of Energy

  As a result of a chemical reaction, energy is transferred out of a battery.

  Electrical circuits provide a means of transferring electrical energy.

  Electrical energy may be transformed into both thermal energy and kinetic energy of a spinning coil.

MATERIALS

For the Procedure

Per student or per pair

- fresh D-cell battery (1.5 volts)

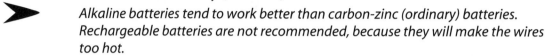

Alkaline batteries tend to work better than carbon-zinc (ordinary) batteries. Rechargeable batteries are not recommended, because they will make the wires too hot.

- short length of dowel rod ⅝–1 inch in diameter or any cylindrical object of similar size (used for winding coil)
- 75–100 cm of 22-gauge varnished (enameled) copper wire (sometimes called magnet wire)

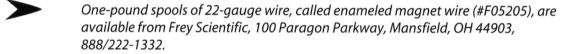

One-pound spools of 22-gauge wire, called enameled magnet wire (#F05205), are available from Frey Scientific, 100 Paragon Parkway, Mansfield, OH 44903, 888/222-1332.

- small piece of fine sandpaper
- thick rubber band
- 30 cm of 14-gauge unvarnished copper wire

Spools of 14-gauge bare copper wire (#74-230-2134) are available from Delta Education, P.O. Box 3000, Nashua, NH 03061-3000, 800/442-5444. You may also be able to purchase solid house-wiring wire from a hardware store. Slit the insulation, remove the two insulated wires, and strip off this extra insulation. These two wires are 14-gauge copper wire.

- clay
- alligator clip wires
- doughnut-shaped ceramic magnet (about 3 cm diameter)

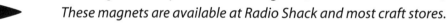

These magnets are available at Radio Shack and most craft stores.

Per class

- pair of needle-nose pliers with cutting edges (or several pairs if students bend wire themselves)
- several 6-volt or 9-volt batteries
- magnets with varying strengths
- additional varnished wire for making coils with different numbers of turns
- insulating material such as paper, plastic, or cloth to fit over the ends of the 1.5-volt battery
- goggles

SAFETY AND DISPOSAL

Wear goggles when cutting the 14-gauge wire. No special disposal procedures are required.

GETTING READY

Build one mini-motor for demonstration. Cut the 14-gauge wire into 15-cm pieces.

If time is limited or if the students lack the ability to bend the 14-gauge wire, you could shape the wire before class. Because of the tendency of the 22-gauge wire to get tangled, it is easier to cut it as you pass it out. The exact length of the 22-gauge wire is not critical, so after you measure one or two you can just approximate.

INTRODUCING THE ACTIVITY

Demonstrate your mini-motor for the class. Ask students where the kinetic energy of the coil comes from. Ask students what aspects of the design (variables) might affect how fast the motor spins. (Possible variables are size of battery, size of coil, number of turns of wire, strength of magnet, kind of wire, and shape of coil.) Tell students they are going to build a motor and test the effects of some of these variables.

PROCEDURE

1. Pass out materials and Mini-Motor Assembly Instructions (provided). Have students assemble their mini-motors.

2. Have students experiment with their mini-motors as directed on the Investigation Sheet (provided). Steps 4–8 on the Investigation Sheet can be done in any order to facilitate sharing of magnets and batteries. The first few pairs of students who do steps 7 and 8 will need to make new coils. Other students can just borrow these. For intermediate students, you may want to assign each pair (or student) to test just one of the five variables in steps 4–8.

3. Discuss the results of these experiments with the class.

4. If each student makes his or her own motor, you may want to assign students to take the motors home and explain the energy transformations to someone else in their families.

EXPLANATION

The following explanation is intended for the teacher's information. Modify the explanation for students as required.

In this activity, students make a simple direct current (DC) motor. An electric motor works by changing the chemical potential energy of a battery into mechanical energy in the form of a rotating coil. The electric motor is a convenient source of motive power because it is clean and silent, starts instantly, and can be built large enough to drive a fast train or small enough to run a tiny wristwatch. Commercial motors may run on either direct or alternating current, may use permanent magnets or electromagnets, and may contain several coils. An electric motor requires three elements: a source of current, a magnet, and one or more loops of wire that are free to turn within the magnetic field of the magnet.

In this mini-motor, the current is provided as a direct current by the battery, the magnet is a permanent ceramic magnet, and the loop consists of coiled copper wire. When the circuit is closed (that is, when the posts at the negative and positive ends of the battery come into contact with the leads from the coil) the battery produces a current in the copper wire. A loop carrying a current in a magnetic field has forces on it that cause it to turn, as shown in Figure 1. If the current flows in one direction without interruption, the force deflects the coil a certain amount in one direction, and the coil stays in that position. If the current begins with the coil vertical, forces are created by the magnet which push it toward a horizontal position. Once it passes horizontal, the effect of the force changes and the coil tries to reverse direction and return to horizontal. The coil may oscillate a few times but will eventually settle in a horizontal position. You may occasionally notice this behavior with the mini-motor, but it is not the desired behavior; the coil of a properly constructed mini-motor should turn continually.

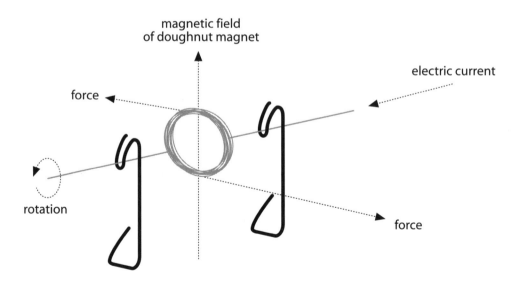

Figure 1: A loop carrying a current in a magnetic field has forces on it that cause it to turn.

So how do you get a motor to continue to turn? In a typical motor, the current does not continually flow through the coil in one direction. Rather, each half-turn of the coil reverses the orientation of the coil with respect to the battery. This reversal causes the flow of the current through the coil to be reversed with each half-turn. As a result, the force keeps changing direction, pushing the coil forward with every half-turn. In an alternating current (AC) motor, the reversal of the current happens automatically. In a DC motor, the coil and its leads must be built in a special way to make the current reverse.

The mini-motor in this activity works slightly differently from the motor just described. In the mini-motor, the flow of the current through the coil is interrupted with each half-turn instead of reversed in direction. When the current is flowing, the force deflects the coil. When the current stops, the force stops pushing the coil, but the coil's inertia carries it through the turn and back to the starting position. When the current comes back on, the coil gets another push. A motor that operates in this way does not have as much force as one with a reversing current because the coil is being pushed only half the time.

In the mini-motor, the current can be interrupted in two ways. One method is to remove the varnish from only one side of each coil lead (for example, the top). With every half-turn of the coil, the varnished side touches the battery leads, breaking the circuit. (See Figure 2.)

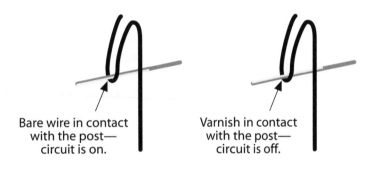

Bare wire in contact
with the post—
circuit is on.

Varnish in contact
with the post—
circuit is off.

Figure 2: A wire with varnish removed from one side of the coil lead will break the circuit every half-turn.

However, in this activity we removed the varnish from the entire coil lead. At first glance it seems that the motor should not work, because the current can flow through the coil leads without interruption no matter what the orientation of the coil. But because the coil bounces as it turns, the coil leads move away from the battery leads and break the connection. (See Figure 3.) If the coil were perfectly stable and the coil leads remained in contact with the battery leads at all times, our method would not work. As the current intermittently starts and stops, on average the force in one direction will be greater than the force in the other. It is not possible to predict in which direction a mini-motor will rotate.

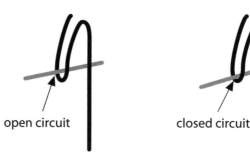

open circuit   closed circuit

*Figure 3: As the coil bounces, the leads move away
from the posts, breaking the circuit.*

Increasing the voltage of the battery or the strength of the permanent magnet will increase the force on the coil, causing it to turn faster. Students will probably notice little effect from changing the shape or size of the coil unless they make the coil very small. Theoretically, using a coil with a larger area will increase the rotational effect of the force (the torque), but it also increases rotational inertia, which makes the coil more difficult to turn. Likewise, adding more turns increases both the force on the coil and the mass of the coil.

Any time electric current moves through a wire, some of the electrical energy is transformed into thermal energy of the atoms in the wire. This energy manifests itself as an increase in temperature. After the motor has been running for a while, students can easily observe that the wire is getting hot. Electric stoves and electric heaters are based on this principle of conversion of electrical energy into thermal energy.

CROSS-CURRICULAR INTEGRATION

Social studies:
*   Have students research and write a report on the history of electric motors or the impact of the invention of the electric motor on society.

REFERENCES

Macaulay, D. *The Way Things Work;* Houghton Mifflin: Boston, MA, 1988.
Renner, A.G. *How to Make and Use Electric Motors;* Putnam: New York, NY, 1974.
Strongin, H. *Science on a Shoestring;* Addison-Wesley: Reading, MA, 1976.

CONTRIBUTOR

Gary Lovely, Edgewood Middle School, Hamilton, OH; Teaching Science with TOYS peer mentor.

HANDOUT MASTERS

Masters are provided for the following handouts:

- Mini-Motor Assembly Instructions
- Investigation Sheet

Copy as needed for classroom use.

Make Your Own Motor

Mini-Motor Assembly Instructions

Some students should start with step 4 to facilitate sharing of pliers.

1. Form the coil by wrapping varnished copper wire tightly around the dowel rod. Leave approximately 6 cm of wire on each end of the coil. These will be the lead wires. (See Figure 1.) The lead wires should be symmetrical—half the circle should be above the leads and half below.

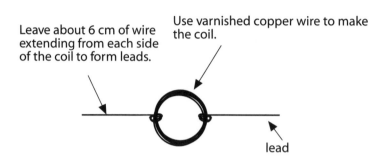

Leave about 6 cm of wire extending from each side of the coil to form leads.

Use varnished copper wire to make the coil.

lead

Figure 1: Form the coil.

2. Wrap each lead around the coil twice, then pass it from the center to the outside through the individual turns of wire.

3. Straighten the lead wires, then sand the varnish off the leads.

4. Using pliers, bend both pieces of 14-gauge bare copper wire into approximately the shape shown in Figure 2. The two pieces should be close to the same height when complete.

Figure 2: Bend the wire into this shape.

Reproducible page from *Exploring Energy with **TOYS*** published by Terrific Science Press™

5. Use the rubber band to attach the bent wires to the ends of the battery. These posts form a stand to hold the coil. (See Figure 3.) Adjust the wires so the loops on the posts are at the same height.

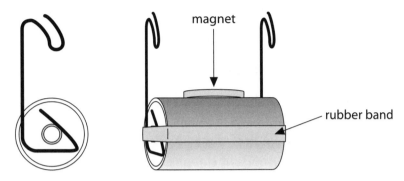

Figure 3: Use a rubber band to attach wires to the ends of the battery, and then place a magnet on top.

6. Place the magnet on top of the battery. (Some battery casings will attract the magnet, so further support is unnecessary. However, for other batteries, you may need to stick the magnet to the casing with a small piece of clay.) Set the battery in a clay base so it doesn't roll over.

7. Place the coil leads in the open loops on the posts. Adjust the height of the loops until the coil is level and rests just above the circular magnet without touching it. Trim any excess wire from the coil leads. (See Figure 4.)

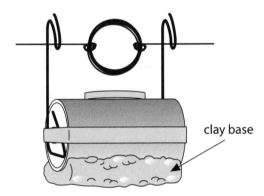

Figure 4: Place the coil in the loops above the battery.

8. Remove the magnet. Push the coil gently to start it spinning. Note how long the coil continues to spin before stopping.

9. Replace the magnet under the coil. Push the coil to start it spinning. Now it should continue to spin much longer. (In fact, it may continue until you stop it.) If your coil doesn't continue to spin longer than it did without the magnet, you need to do some troubleshooting. Try making some of the following adjustments:

    •   Make sure the coil is level.

- Make sure the coil leads are placed symmetrically—half the circle should be above the leads and half below.

- Make sure the part of the leads that rests on the posts is well sanded.

- Bend the posts to bring the coil closer to the magnet, making sure to keep the coil level.

- Tilt the magnet to one side so it is not perfectly centered on top of the battery.

- If the coil slides toward one end of the battery while rotating or oscillating, bend the post on that end to raise the loop slightly. (Keep in mind that the coil does not need to be centered between the posts. In fact, the mini-motor may work very well if it and the magnet are closer to one post than the other. The important thing is to keep the coil from sliding as it rotates.)

- Move the coil a bit to the left or right of the center of the magnet. Keep moving the coil until you find an off-center spot that works.

- Add an extra magnet to the motor, or replace your magnet with a stronger one. If necessary, adjust the height of the posts.

- If nothing else works, have your teacher check your battery. You may need a new one.

If these adjustments still don't get your motor working, consult your teacher for assistance.

Name _____   Date _____

Make Your Own Motor

Investigation Sheet

1. After your motor has been spinning for a while, feel the coil and posts. Are they warm? Give another example of wires getting hot when an electric current passes through them.

    _____

    _____

2. Describe the energy transformations that take place as the motor runs.

    _____

    _____

    _____

3. Give an example of a common device that uses an electric motor to make something spin.

    _____

    _____

4. Replace your magnet with one that is stronger or weaker. Describe the effect (if any) this has on your motor.

    _____

    _____

    _____

5. Return to your original magnet and try using a higher-voltage battery. The easiest way to do this is to slip a piece of insulating material such as paper, plastic, or cloth between your bent wires and the battery. Then use alligator clip wires to attach the new battery to the posts. Describe the effect (if any) this has on your motor.

    _____

    _____

    _____

6. Return to your original battery. Carefully change the shape of your coil into an oval or square. Describe the effect (if any) this has on your motor.

_____

_____

_____

_____

7. Make or borrow a new coil with more or fewer turns. Describe the effect (if any) this has on your motor.

_____

_____

_____

_____

8. Make or borrow a coil with a larger or smaller area. Describe the effect (if any) this has on your motor.

_____

_____

_____

_____

9. Summarize the results of your investigations.

_____

_____

_____

_____

_____

_____

Chemical Energy Transformations

... Students observe common toys and household items that convert chemical energy to light, heat, and sound energy.

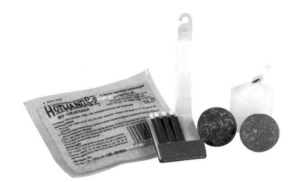

Common items that use chemical energy

✔ Time required

Setup   5 minutes
Performance   30 minutes
Cleanup   5 minutes

✔ Key Science Topics

- chemical energy
- energy transformations
- light energy
- potential energy
- sound energy
- thermal energy

✔ Student Background

Students should already have been introduced to kinetic energy, gravitational potential energy, and energy transformations.

✔ National Science Education Standards

Science as Inquiry Standards:

- Abilities Necessary to Do Scientific Inquiry

  Students observe the production of heat, light, and sound by chemical reactions.

Physical Science Standards:

- Transfer of Energy

  Energy can be transferred out of a system by a chemical reaction.

MATERIALS

For Introducing the Activity
Per class
- any object that will neither bounce nor break when dropped
- magnet
- paper clip
- 2 balloons
- 2 pieces of string 18–24 inches long
- overhead projector
- (optional) piece of silk cloth

For the Procedure
Per class
- 2 lightsticks
- clear plastic cup made of high-density polyethylene (HDPE)

➤ *Do not use polystyrene or Styrofoam™, as the liquid inside the lightstick may dissolve these materials.*

- (optional) 6% hydrogen peroxide solution (available at beauty supply stores)
- utility knife or pocket knife
- set of Blaster Balls (also called Crackling Balls)

➤ *If you are not able to find these balls in a toy store, you can order the same toy, called Hand Blasters (#10662) from American Science and Surplus, 3605 Howard St., Skokie, IL 60076, 847/982-0870.*

- 1 of the following
  ○ several chemical hand warmer packets and a disposable plastic plate

➤ *Chemical hand warmers are sold under a variety of names. The two most common ones are Heat Packs by R. G. Barry Co. and Hot Hands by Heatmax Manufacturing, Inc. Look for them in sporting goods stores or department stores during fall and winter.*

  ○ HeaterMeal by HeaterMeals, Inc.

➤ *HeaterMeals are available in the packaged foods section (not frozen) of grocery stores and truck stops.*

- candle
- matches or lighter for lighting candle
- goggles

For the Extensions
❶ Per student or group
- lightstick
- container of hot tap water
- container of ice water

❷ Per student or group
- concentrated powdered detergent
- water
- small cup or zipper-type plastic bag

SAFETY AND DISPOSAL

This activity is intended to be done as a demonstration. The following safety precautions are for you, the teacher.

Goggles should be worn while dissecting the lightstick. If you have very sensitive skin, you may want to wear latex gloves while working with the chemicals in the lightstick. For most people, simply washing hands afterwards should prevent skin irritation. Do not use the HeaterMeal near a fire or flame (including the candle) because it releases flammable gas.

Used chemicals from the lightstick can be washed down the drain with water. The lightstick tube should be placed in a trash can away from children as it will contain small slivers of glass. The knife and the plastic cup will not be food-safe after touching the lightstick chemicals; either make sure they are never used with food or drink again or throw them away. Set the hand warmers and HeaterMeal aside until you are sure they are no longer producing heat. They can then be placed in a waste basket.

GETTING READY

If using hand warmer packets, activate several of them just before the lesson begins. Some brands have a few minutes delay before they start producing heat, so if you start them before the lesson they should be ready when you want to pass them around.

Blow up two balloons and tie a string to each balloon.

INTRODUCING THE ACTIVITY

Review the concept of potential energy with the students. Hold up an object that will neither break nor bounce when dropped. Ask the students what kind of energy it has now. *Gravitational potential energy.* Release the object and allow it to fall. Ask the students what kind of energy it has just before it hits the floor. *Kinetic energy.*

Place a magnet and a paper clip on an overhead projector. Position the paper clip so that, when you release it, it will be pulled to the magnet. Release the paper clip, then ask the students how the result is similar to what happened when the object was dropped. *The clip gained kinetic energy when released, just as the dropped object did.* Point out that the clip must have had some kind of potential energy before it was released.

Charge the two inflated balloons by rubbing them with a silk cloth or against your hair. Have one student hold the strings attached to the two balloons so that they are about a foot apart. Have other students hold each balloon (touching it as little as possible) so that it hangs straight down. Then have the students simultaneously release the balloons, which will move away from each other. Again ask the class how this is similar to the dropped object. Ask the students to identify the force responsible for the potential energy in each case. *Gravity, magnetic force and electrical force.*

PROCEDURE

Explain to the students that the form of energy they have called chemical energy while studying energy transformations is really a form of electrical potential energy similar to the energy the balloons had before they were released. Atoms and molecules have a certain amount of electrical potential energy because of the relative positions of the various charged particles of which they are made and the forces these particles exert on one another. When atoms or molecules undergo a chemical reaction, the atoms or molecules of the substances produced in the reaction may have more or less electrical potential energy than the original atoms or molecules. If they have less potential energy, then the lost potential energy has been transformed into some other kind of energy, such as thermal energy or light. If the product substances have more potential energy, then energy must have been added to the molecules. Tell the students that you are going to demonstrate several different chemical reactions in which chemical energy is transformed into some other kind of energy. In each case, they should identify the kind of energy that is produced.

Chemical Energy to Light: Lightstick

1. Hold up a lightstick. Ask the students to raise their hand (without saying anything) if they know what this is.

2. Activate the lightstick by bending it. Ask for hands again if there were some students who didn't initially know what it was. Call on someone to identify what kind of energy the chemical energy has been transformed into.

3. Ask if anyone knows why the lightstick was not glowing initially but then started glowing after it was activated. *(Breaking the inner glass tube allowed two substances to mix with each other, causing a chemical reaction to take place.)*

4. Remove a new lightstick from its packet and tip it so that the glass tube inside slides to the bottom of the lightstick. Wearing goggles, use a knife to cut about halfway through the lightstick near the top of the tube.

5. Pour the liquid out into a clear plastic cup, without allowing the small glass tube to come out.

6. Ask a student to describe the liquid in the cup. Make sure he or she points out that it isn't glowing.

Be very careful to hold the end of the lightstick closed if breaking the glass tube. The glass tube is made of very thin glass which can easily cut your hand or fly out of the lightstick when broken. Alternatively, instead of breaking the glass tube, you can use 3–4 mL 6% hydrogen peroxide solution in step 8 and omit step 7.

7. If using the hydrogen peroxide from the glass tube, hold the lightstick over the cup. Making sure the glass tube is near the middle of the lightstick and holding the cut end closed, bend the lightstick to break the tube.

8. Pour the second liquid (either the solution from the glass tube or 3–4 mL 6% hydrogen peroxide solution) into the cup.

9. Swirl the cup to mix the liquids.

10. Now ask a student to describe the resulting liquid. Reiterate that, as the molecules of the two chemicals interact, some of their component charged particles change position and lose electric potential energy. This released energy appears in the form of light energy.

Explaining exactly how this light is produced requires an understanding of atoms at a level significantly beyond that of middle school students.

Chemical Energy to Sound: Blaster Balls

1. Hold up the Blaster Balls. Ask if anyone has ever used this toy. If so, ask them to explain how the Blaster Balls are used before you demonstrate them.

2. Demonstrate the toy by striking the balls together sharply. Ask the students what kind of energy is being produced. *Sound energy.*

3. Give the Blaster Balls to one or two students to examine. Ask them what indications show that a chemical reaction has taken place. *An odor of burning is detectable, and dots of a different color appear where the balls touched.*

If desired, make the Blaster Balls available to students later in the day or during recess.

4. Explain to the students that in this case just putting the two chemicals (one from each Blaster Ball) in contact is not enough. Some extra energy must be added to get the reaction to begin. Kinetic energy of the moving balls is converted to thermal energy when the balls collide. This allows the reaction to begin.

5. Ask the students what other toys they can think of that behave like the Blaster Balls. *(Cap guns and the small firecrackers known as Snappers or Pop-its are good examples.)*

Chemical Energy to Heat

Complete one of the two demonstrations on conversion of chemical energy into thermal energy described below.

Hand Heater

1. Show the hand warmer package to the students and ask whether any students have ever used these packets to keep their hands warm while outdoors.

2. Open the outer package. Allow one or two students to touch the inner packet and report its temperature.

3. Shake the inner packet. It should immediately become warm. Allow the same students to touch it and report to the class that it is now warm.

4. Ask the students what kind of energy is being produced. *Thermal energy.* Pass around the packets you activated before starting the lesson.

5. Cut open one of the bags and dump the contents out on a disposable plastic plate. Walk around the room and let the students see the contents and feel the bottom of the plate.

6. Tell the students that some of the substances in the bag are reacting with the oxygen in the air. In this process, some of their electric potential energy is being converted into thermal energy.

HeaterMeal

1. Ask the students if they have ever eaten a "HeaterMeal." These products are fairly new on the market, so many of the students may not have heard of them before. If so, explain that these are packaged meals that are stored at room temperature and have a built-in heat source so they can be eaten hot. Thus, they can be very convenient when you want to have a hot meal but are not near a kitchen or other cooking facilities.

2. Open the box and allow the students to see the chemical packet attached to the bottom of the Styrofoam tray. (This packet is much like a hand warmer packet.) Allow someone to feel the packet, and ask whether it is hot.

3. Open the packet of water and pour it over the chemical packet in the tray. The water will quickly start to boil and give off steam and hydrogen gas. Thermal energy has been produced.

Do not touch the steaming tray of the HeaterMeal or allow students to touch it; it is extremely hot. Do not use near fire or flame, because hydrogen, a flammable gas, is released in the reaction.

4. Allow the students to view the steaming tray without touching it.

Chemical Energy to Light and Heat: Candle

If you perform the HeaterMeal demonstration before this one, either open a window or wait a few minutes before lighting the candle to allow any hydrogen gas produced by the HeaterMeal to dissipate.

1. Light a candle. Ask students what kind of energy is being produced. *Light and thermal energy.*

2. Ask students whether they know how energy is produced when something burns. If no one does, explain that some of the molecules of the burning material (usually ones containing the element carbon) undergo a chemical reaction with the oxygen in the air to produce other substances. (For carbon-containing substances, the primary products of this reaction are usually water and carbon dioxide.) As a result of this chemical reaction, electrical potential energy is changed to thermal energy and light. In this case, it is the candle wax that is burning (or reacting). Point out to the students that some of the wax has changed its physical state from solid into liquid. Some of the candle wax has also changed into the gas state. It is this wax vapor that burns.

CLASS DISCUSSION

Review the lesson with the students. The lightstick produced light energy. The Blaster Balls produced sound energy. The hand warmer or HeaterMeal produced thermal energy. The candle produced both thermal energy and light. In all cases, atoms or molecules lost electric potential energy when their constituent parts were rearranged in a chemical reaction. This potential energy was transformed into other forms of energy.

Ask the students to brainstorm some other devices or materials in which a chemical reaction causes an energy transformation. Other examples are fireworks, which produce light, sound, and thermal energy, and fireflies which produce light energy in a manner similar to the way lightsticks do. The processes by which our bodies produce energy from the food we eat are similar to the burning candle in that the products are carbon dioxide and water. In these biological processes only thermal energy is produced, not light as with the candle. Students may remember working with Plaster of Paris and know that it can become quite warm after the powder is mixed with water. This thermal energy is produced by a chemical reaction.

EXPLANATION

The following explanation is intended for the teacher's information. Modify the explanation for students as required.

Atoms, molecules, and ions have electrical potential energy resulting from the relative positions of their electrons and protons. When these components are rearranged during a chemical reaction, they either gain or lose electrical potential energy. If they gain electrical potential energy, then energy must have been added to the system in order for the reaction to take place. If they lose electrical potential energy, then that energy lost from the system is transformed into another form. Most frequently the energy released is in the form of thermal energy, as is seen with the hand warmers and HeaterMeals.

In the case of Blaster Balls, we have indicated in the lesson that sound energy is produced, which is certainly the case. However, chemical energy is not transformed directly into sound energy. First, heat and a gas are produced by the chemical reaction. This hot gas rapidly expands, producing sound in much the same way that thunder is produced when the rapidly moving electric charges in a lightning bolt heat the air. Some light is also produced by the Blaster Balls, but it is visible only in a completely darkened room. Even then it can be hard to see, so we do not suggest that you attempt to show it to the students.

Most common sources of light also produce heat. For example, the common incandescent light bulb releases much more energy in the form of heat than of light. However, a small number of chemical reactions produce light without heat. This phenomenon is called chemiluminescence. The lightstick contains a dye and other substances including an ester. The little glass tube contains a hydrogen peroxide solution. When you break the glass tube, the ester and the hydrogen peroxide react. This process causes a transfer of energy to the dye molecules, which then release it in the form of light. The identity of the dye determines the color of the light produced.

The hand warmers contain iron that reacts with oxygen in the air. The iron is powdered to produce a greater surface area, increasing the number of iron atoms that will come into contact with oxygen molecules. You shake the packet in order to mix more air into the powder. This reaction proceeds slowly and the hand warmer will stay warm for hours. The food heater in the HeaterMeal is similar to the hand warmer, but it contains magnesium as well as iron. The chemical reaction occurs much more rapidly and produces much more thermal energy—enough thermal energy to raise the temperature of the water to the boiling point. In less than an hour, the reaction will cease and the packet will cool off.

The burning of the candle is a typical combustion reaction, in which compounds containing the elements carbon and hydrogen react with oxygen to form the compounds water and carbon dioxide. The released energy appears in the form of thermal energy and light.

EXTENSIONS

1. The rate at which the chemical reaction in the lightstick occurs is dependent on temperature. A lightstick placed in hot tap water will glow more brightly, while one placed in ice water will glow more dimly. Allow the students to investigate this change in the rate at which energy is being converted. Further information and complete lesson plans can be found in Sarquis et al. 1995. (See References.)

2. Mix concentrated powdered laundry detergent with a small amount of water in a zipper-type plastic bag or small cup. A definite increase in temperature is noticeable as the water is stirred into the detergent. (Some debate exists among chemists as to whether dissolving should be considered a chemical change or a physical change like melting. However, in either case, energy is changing form.)

FURTHER READING

Craig, A; Rosney, C. *The Usborne Science Encyclopedia;* EDC Publishing: Tulsa, OK, 1988. (Students)

REFERENCES

"Chemiluminescence: Dissecting a Lightstick"; *Fun with Chemistry: A Guidebook of K–12 Activities;* Sarquis, M., Sarquis, J., Eds.; Institute for Chemical Education: Madison, WI, 1991; Vol. 1, pp 163–166.

"Energy Changes with Everyday Materials"; *Fun with Chemistry: A Guidebook of K–12 Activities;* Sarquis, M., Sarquis, J., Eds.; Institute for Chemical Education: Madison, WI, 1991; Vol. 2, pp 223–228.

Sarquis, M., Sarquis, J.L., Williams, J.P. "Investigating the Effect of Temperature on Lightsticks"; *Teaching Chemistry with Toys: Activities for Grades K–9;* Terrific Science: Middletown, OH, 1995; pp 269–274.

Concept Introduction

✔ **Concept Application**

Synthesis

Simple Machines with LEGO

...Students use fun and popular LEGOs to build working simple machines.

✔ Time Required

Setup   5 minutes
Performance   40 minutes (for steps 1–3)*
Cleanup   5 minutes

*The rest of the activity is open-ended and will require several hours spread out over several days.

Lever built with LEGOs

✔ Key Science Topics

- definition of work
- simple machines

✔ Student Background

Students should be familiar with levers, pulleys, gears, and wheels. At least some students in each group should have prior experience building with LEGOs®. LEGO Educational sells some excellent curricular materials on simple machines that could be used to provide this background if needed. (See Introducing the Activity.)

✔ National Science Education Standards

Science as Inquiry Standards:
- Abilities Necessary to Do Scientific Inquiry
  Students make models of simple machines.
  Students communicate their results orally and with drawings.

Physical Science Standards:
- Transfer of Energy
  Energy can be transferred by simple machines.

✔ Additional Process Skill

- classifying   Students classify LEGO models according to the type of simple machine represented.

MATERIALS

For Introducing the Activity
Per class
- pictures of familiar objects that incorporate or are simple machines (See Introducing the Activity.)

For the Procedure
Per group
- LEGO Technic Universal Set #8032 or any larger LEGO Technic set
- string for pulleys
- (optional) pennies or paper clips

SAFETY AND DISPOSAL

No special safety or disposal procedures are required.

INTRODUCING THE ACTIVITY

Review levers, pulleys, wheels, and gears by showing pictures of familiar objects that incorporate or are simple machines and discussing how they work. Pictures might include a seesaw, wheelbarrow, truck loading ramp, and bicycle. You do not need to review screws and inclined planes, because they can't be built effectively with LEGOs. Alternatively, excellent curriculum materials that you could use to introduce the students to simple machines are available from LEGO DACTA, 555 Taylor Rd., P.O. Box 1600, Enfield, CT 06083-1600, 800/527-8339.

PROCEDURE

If sufficient LEGOs are available, steps 1 and 2 are best done in small groups with the entire class working at once. Steps 4–5, which are more complex, can be done in small groups with the entire class working at once or by individuals or groups working in a Learning Center over several days.

1. Challenge each group to assemble examples of as many simple machines as they can from their LEGOs. At minimum, they should build a lever, construct something with a working pulley in it, construct something in which at least two gears mesh, and construct something that incorporates a wheel and axle. Encourage students to make several different kinds of levers. (See Figure 1 for sample levers.) Circulate through the class with a checklist, marking as each group finishes each type of model.

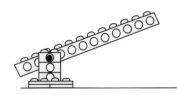

a. simple seesaw-type lever

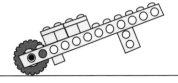

b. lever pivoted at one end, as in a wheelbarrow

c. more complex system including several levers

Figure 1: Sample levers built with Legos

2. If students have not had hands-on experience with levers, they should test the lifting ability of their levers by putting weights such as pennies or paper clips on the load end and then stacking the same kind of weights on the other end until the load is lifted. They could also investigate the effect of changing the position of the fulcrum (pivot point). They will see that if the fulcrum is closer to the load end, the load can be lifted by a smaller weight.

This works well as a free exploration, but if you feel your students need more structure, use the Investigation Sheet (provided).

3. Discuss why simple machines are used. (See the Explanation.)

4. Tell students that they will be designing a special model using LEGOs. Explain that this object should demonstrate as many simple machines as possible, all in the same model. The simple machines can be the same or different types. For several days, provide opportunities for individual students or teams of two or three (but no more) to go to a table where at least one box of LEGOs is kept and work on a design. They should make notes and sketch any ideas they come up with.

Note-taking and sketches are important because the next students to work with the LEGOs will probably take apart whatever the previous group built. Without a record of their ideas, students will forget between building sessions.

5. At the end of the designated time, instruct students to turn in their drawings and notes. Remind students that the drawings must be clear and the descriptions easy to understand. Post the drawings and give students time to examine them. Then have each student or group make a brief presentation using a sketch of the design. After everyone has presented, instruct the class to decide by vote which idea to use. Have the winning designer(s) build the plan into reality and set it out for display, along with a poster that lists all the simple machines in the model.

EXPLANATION

The following explanation is intended for the teacher's information. Modify the explanation for students as required.

Contrary to statements found in many textbooks, simple machines do not allow us to do less work. If we were able to accomplish the same results with less work, we would be able to output more energy than we put in. No simple machine allows us to do this. In fact, due to friction between moving parts, simple machines usually require more work to be input than they give in output.

If simple machines require more work than they put out, why do we use them? In most cases, they allow us to accomplish the same result (lifting a heavy box, for instance) using a smaller force that is easier for our muscles to generate. However, this advantage requires a trade-off. If less force is used, we must move the force over a greater distance. Why? Because the work done (the product of force and distance) stays the same, in order to use a smaller force we must simultaneously increase distance. For example, imagine lifting a 100-newton box up into the back of a truck 1 m above the ground. We can lift it straight up using a 100-newton lifting force and doing 100 joules of work, or we could slide it up a 4-m-long inclined plane using a 25-newton force and still doing 100 joules of work. (Actually, if we were really moving the box up the inclined plane, we would need a force somewhat larger than 25 newtons because of the friction between the inclined plane and the box.) The same principle applies to doing work with other simple machines such as levers, gears, and multiple-pulley systems. The applied force required is smaller, but it must move a greater distance.

Another reason simple machines are used is to make something move faster than we can easily do by applying a force directly to it. An example of this use is a hand-operated drill. The bit must turn much faster than you can easily move your hand. A progression of larger to smaller gears enables the bit to turn much faster than your hand is moving. The trade-off is that your hand must move in a much larger circle (travel a greater distance) than if you were turning the bit directly with your hand.

CROSS-CURRICULAR INTEGRATION
Art:
*   The drawings for the designs can be a good lesson in art as a form of accurate communication. Drawings must be clear and follow rules of proper proportion and perspective.

Language arts:
- The written descriptions of the design plans should be clear, coherent, and appropriately organized. Have students use peer editors for user testing and proofreading.
- Read aloud or suggest that students read the following story:
  - "The Big Parade" in *Einstein Anderson Goes to Bat,* by Seymour Simon (Puffin, ISBN 0-14-032303-1)
    Einstein uses his understanding of levers to trick the school bullies into carrying more than their share of a heavy load.

FURTHER READING

Cooper, C.; Osman, T. *How Everyday Things Work;* Facts on File: New York, NY, 1984. (Teachers)

Taylor, B. *Get It in Gear: The Science of Movement;* Random House: New York, NY, 1991. (Students)

Walpole, B. *Fun With Science: Movement;* Warwick: New York, NY, 1987. (Students)

CONTRIBUTOR

Anita Kroger, Gifted and Talented Specialist, Cincinnati, OH; Teaching Science with TOYS, 1986–87.

HANDOUT MASTER

A master for the following handout is provided:
- Investigation Sheet

Copy as needed for classroom use.

Name _____ Date _____

Simple Machines with LEGO
Investigation Sheet

1. Build a simple lever with the fulcrum (pivot point) in the exact center (like a seesaw). Using pennies or paper clips as weights (or your smallest LEGO blocks), prove to yourself that you can always balance the lever with equal weights on the ends.

2. Now put one object on one end of your lever and determine where you must put two identical objects on the other end for it to balance. (Working with a partner makes this easier. One of you can keep the one object from falling off while the other moves the two objects.) When you have balanced the lever, use a ruler to estimate the distance from the center of each object to the fulcrum.

    Distance from two objects to fulcrum _____

    Distance from one object to fulcrum _____

    Compare these two numbers. Which one is bigger? _____

    What is the approximate relationship between the two distances? (Hint: consider ratio.) _____
    _____

    Compare the numbers of objects on each side of the lever. What is the relationship (ratio) between those two numbers? _____

3. Repeat step 2 using three objects on one side and one object on the other.

    Distance from three objects to fulcrum _____

    Distance from one object to fulcrum _____

    Compare these two numbers. Which one is bigger? _____

    What is the approximate relationship between the two distances? (Hint: consider ratio.) _____
    _____

    Is this the same answer you found in step 2? If not, why might they be different?

    _____

    _____

    _____

Reproducible page from *Exploring Energy with **TOYS*** published by Terrific Science Press™

4. Carefully move the lever up and down a little. Does the single penny move up and down a greater distance than the three pennies? (Don't try to measure these distances, because it is hard to measure such small distances with a ruler. However, if we could measure them, we would find that the single penny moves about 3 times as far as the three pennies.)

5. Now move the fulcrum to another position of your choosing. Put two pennies on the short end and determine where you need to put one penny to make it balance. Measure your distances again.

   Distance from two pennies to fulcrum _____

   Distance from one penny to fulcrum _____

   What is the approximate ratio between these two numbers? _____

   Why do you think this number is different from the relationship (ratio) in step 2?

   _____

   _____

   _____

   _____

   _____

Get It in Gear with a LEGO Vehicle

...Students explore the transmission of energy using gears.

✔ Time Required

Setup   10 minutes
Performance   60 minutes*
Cleanup   10 minutes

* Two 30-minute classes. The time will vary depending on the amount of prior experience your students have with LEGOs®.

✔ Key Science Topics

- forces
- gears
- transfer of energy
- work

A LEGO Gyro-Copter

✔ Student Background

Students should be familiar with the basic types of simple machines and the concept of work.

✔ National Science Education Standards

Science as Inquiry Standards:

- Abilities Necessary to Do Scientific Inquiry

  Students observe and describe the motion of gears.

  Students feel the force required to move the rotor in different ways.

  Students base explanations on their observations and experience.

Physical Science Standards:

- Transfer of Energy

  Energy is associated with mechanical motion and can be transferred by gears.

MATERIALS

For Getting Ready
Per class
- (optional) LEGO Technic #8215 or other LEGO Technic set for building a sample

For the Procedure
Per pair of students
- zipper-type plastic bag for storing parts
- washable marker
- LEGO Technic #8215 or other LEGO Technic set

 Most LEGO Technic sets will have at least one model in the instruction booklet that would be suitable for the activity. If a model other than #8215 is used, the Investigation Sheet (provided) will need to be revised as described in Getting Ready. Groups can use different sets of LEGOs. This will mean more work for you in preparing Investigation Sheets, but it does facilitate having students bring in sets from home rather than purchasing them. In this case, you should have each group present their model to the class and explain how it works.

SAFETY AND DISPOSAL

No special safety or disposal procedures are required.

GETTING READY

You may want to build a sample vehicle before class. Having a completed model to refer to can speed the construction process.

Any LEGO Technic model that has at least one pair of different-sized gears that mesh can be used for discussing gear ratios. If you also want to discuss using gears to change the size of the applied force, you need two unequal-sized gears (or wheels) mounted on the same shaft.

If using a set other than #8215, select a model from the instruction booklet and modify the Investigation Sheet (provided) appropriately. If appropriate for your model, you may want to ask the students to identify any other simple machines, such as levers, in the vehicle. (The Gyro-Copter has no other simple machines, so that question is not on the Investigation Sheet.)

PROCEDURE

1. Have students work in pairs to complete the Investigation Sheet. Circulate to provide assistance while they work. You have two options for pairing students for this activity. You may want to pair students who have little LEGO experience with ones who are more experienced. However, in some cases, this can lead to one student doing the

construction while the other watches. Another option is to put your expert builders together and your inexperienced builders together. The inexperienced students will take longer but will gain experience and confidence from having done it on their own. (This option will, of course, require more class time than the first option.)

2. After all pairs have completed the Investigation Sheet, review it with the class, calling on a different pair for each answer. Be sure to discuss why the general rule that relates gear ratios to gear speeds must be true. As you discuss the reason for using two gears mounted on the same axle, ask the students questions to lead them into thinking in terms of forces and energy if this does not arise naturally in the discussion. You might want to let someone bring in a bike or take the class out to the bike rack to look at gears. Have some students describe their experiences with multispeed bikes. Relate this to what they have learned about using gears to change speeds and force.

The Gyro-Copter does not incorporate two gears mounted on the same axle, but it does incorporate rotating objects of different circumferences mounted on the same axle. The crank is mounted on the same axle as the largest gear, and the propeller is mounted on the same axle as the small gear that meshes with the largest gear. The crank and the largest gear are about the same size, so force is not changed between them. However, the small gear is much smaller than the propeller, so the amount of force is changed between them.

EXPLANATION

The following explanation is intended for the teacher's information. Modify the explanation for students as required.

During this activity, your students are able to observe some properties of gears, which are simple machines. A gear is a toothed wheel that precisely meshes with another toothed wheel. Two or more gears are used to change the size of an applied force, the speed of one or more of the gears, or the direction of motion.

Gears can be connected to one another in two ways. They can either mesh or be connected by a chain. (See Figure 1.) Two gears that mesh turn in opposite directions. Two gears that are connected by a chain turn in the same direction.

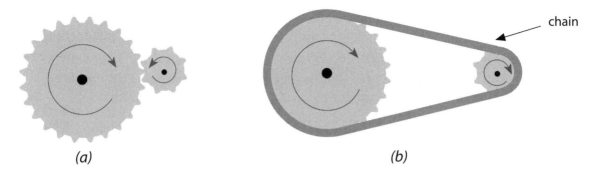

Figure 1: Gears can (a) mesh with one another or (b) be connected by a chain.

Gears of different sizes can be used to change the speed at which some component of the machine (such as the axle) is turning. A small gear connected to a larger gear will turn faster than the larger gear. This is because the total distance traveled by a tooth on the small gear must be the same as the distance traveled by a tooth on the large one. If the circumference of one circle is three times bigger than the other, then the smaller gear must turn three times around for every complete turn of the larger one. In one revolution, the 24-toothed gear will turn the eight-toothed gear three times. (See Figure 2.) In general, the ratio of the number of teeth of the two gears is inverse to the ratio of their angular speeds (angular speed is the number of revolutions the gear makes per second).

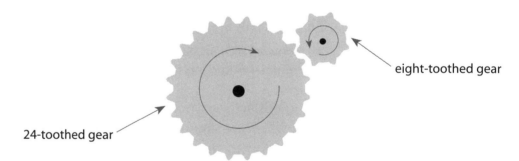

Figure 2: One turn of the 24-toothed gear will turn the eight-toothed gear three times.

As mentioned previously, one of the functions of a gear is to change the size of an applied force. This change can be accomplished when two gears with different circumferences are mounted on the same shaft. (See Figure 3.) In this case, they turn at the same speed, but the teeth travel different distances and apply different amounts of force. When a force is applied to one gear to make the shaft turn, the other gear can be used to apply a force to something else. Since energy input must equal energy output (ignoring losses to friction), the size of the force can be increased or decreased in proportion to the circumferences of the gears.

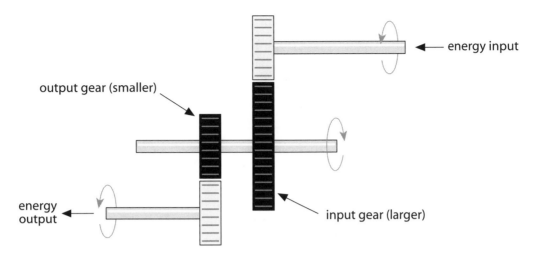

Figure 3: If the input gear is larger than the output gear, then the force is increased. The center pair of gears is used to increase the force applied to the bottom gear. The top gear applies a force to the input gear. Because the output gear is smaller than the input gear, a larger force is applied to the bottom gear.

Here's how gears change force: Work done equals force times distance. The input energy in one revolution is the applied force times the circumference of the gear to which it is applied. The output energy is the circumference of the output gear multiplied by the force applied by it to something else. If the input gear is larger than the output gear, then the output force is increased. If the input gear is smaller than the output gear, the output force is decreased. (The work done—force times circumference—is the same for each gear.) Just as with speed changes, the number of teeth can be used to find the ratio of the forces. If the input gear has 24 teeth and the output gear has 12 teeth, then the output force will be twice as large as the input force.

In the last part of the Investigation Sheet, students are asked to investigate two ways of turning the top rotor. They discover that a larger force is required to turn the rotor with the crank than by spinning the rear propeller. The propeller is mounted on the same shaft as a small gear that meshes with a larger, curved-tooth gear. Because your finger travels in a large circle as it pushes the end of the propeller, the small gear will exert a much larger force on the large gear than the one you are exerting on the propeller. The crank is about the same size as the large gear and is mounted on the same axle, so when you turn the crank, you turn the large gear directly with no multiplication of your force. Another way to think about this is to consider that to achieve a given turning speed for the rotor, you must do the same amount of work whether you turn the crank or the propeller. The circumference of the crank is smaller than the circumference of the circle made by your finger turning the propeller. Thus, in order to do the same amount of work, you must apply a larger force to the crank.

It is difficult to make the rotor turn rapidly by turning the propeller for the same reason that you are able to use a smaller force. Since the large gear and small gear have a 3:1 ratio, the small gear must turn three times in order for the large one to turn once. Thus, in order to achieve the same speed for the rotor as when using the crank, your hand must complete three times as many circles in the same amount of time. This is made even more difficult by the fact that the circles your hand moves in when turning the propeller from the tip are larger than when moving the crank.

In Question 6 on the Investigation Sheet, students are asked to identify a gear that serves no purpose as a gear. This gear is on the upper shaft, which turns the rotor blades, and does not mesh with any other gear. Its function is to hold the axle in the proper position for the gears below it to mesh. Many LEGO Technic models use a gear for this purpose.

EXTENSION

Challenge students to modify the way their vehicles work. Each group should then explain either orally or in writing what their change is and how it works.

CROSS-CURRICULAR INTEGRATION

Social studies:
- Study the invention of labor-saving machines. Students may enjoy reading *Steven Caney's Invention Book* (Workman, ISBN 0894800760).

FURTHER READING

Cooper, C.; Osman, T. *How Everyday Things Work;* Facts on File: New York, NY, 1984. (Teachers)

Teacher's Guide to Technic; LEGO Systems: Enfield, CT, 1985. (Teachers)

Walpole, B. *Fun With Science: Movement;* Warwick: New York, NY, 1987. (Students)

CONTRIBUTOR

Vivian Schulter, Oak Hills School District, Cincinnati, OH; Teaching Science with TOYS, 1986–87.

HANDOUT MASTER

A master is provided for the following handout:
- Gyro-Copter Investigation Sheet for LEGO Technic Set #8215

Copy as needed for classroom use.

Name _____   Date _____

Get It in Gear with a LEGO Vehicle
Gyro-Copter Investigation Sheet for LEGO Technic Set #8215

1. Before you begin construction, color one tooth each on your largest and smallest gears, using a washable marker. Count the number of teeth on each gear.

   large gear _____   small gear _____

2. Open the instructions that came with the set and turn to page 3. Begin construction by completing LEGO Steps 1–10 for the Gyro-Copter (the first model). In Step 10, make sure that the teeth of the small gear mesh well with the curved teeth of the large gear. Adjust the small gear's position slightly if necessary.

3. Continue with LEGO Steps 11–13. After you have finished Step 13, turn the propeller slowly. Use the marked teeth as reference points to count how many times the small gear turns while the large gear turns through one complete circle. How could you have figured this out ahead of time? Devise a general rule that will work for all gears.

   _____

   _____

   _____

4. Complete the model. You may skip LEGO Step 14 if you wish, as it just adds decoration.

5. Turn the crank on the side of the Gyro-Copter. All of the gears should turn. Describe what makes the rear propeller turn. Describe what makes the top rotor turn.

   _____

   _____

   _____

6. Look carefully at all the gears. One of the gears in this toy does not really function as a gear. Which one is it? What is its purpose? How do you know it is not really used as a gear?

   _____

   _____

   _____

   _____

7. Place your finger near the tip of the rear propeller and slowly rotate it. Pay attention to how hard you have to push. Now turn the crank so that the top rotor turns at about the same speed as when you turned the propeller. How does the force you are exerting on the crank compare to the force you exerted on the rear propeller?

_____

_____

_____

_____

8. Why do you think these forces are different? Think about what makes the gear with the curved teeth move in each case.

_____

_____

_____

_____

9. How does the distance your hand moves when you turn the crank compare to the distance your hand moves when you turn the rear propeller?

_____

_____

_____

_____

10. Turn the crank rapidly and watch the top rotor. Now try to make the rotor turn at the same speed by turning the rear propeller as before. Why is this difficult?

_____

_____

_____

_____

_____

_____

Reproducible page from *Exploring Energy with **TOYS*** published by Terrific Science Press™

Squish 'em, Squash 'em, Squoosh 'em

...Students explore the mechanism of the game Grape Escape to review transfer of energy using simple machines.

A Grape Escape game

✔ Time Required

Setup   5–20 minutes*
Performance 45–55 minutes*
Cleanup   10 minutes

*Depending on whether the games are assembled by the teacher or the students.

✔ Key Science Topics

- elastic potential energy
- gears
- transfer of energy
- work

✔ Student Background

Students should be familiar with different types of simple machines and how they work. An understanding of basic energy concepts such as work and elastic potential energy is essential. Students should have completed at least the activities "What Makes It Go?" and either "Pop Can Speedster" or "Toy That Returns."

✔ National Science Education Standards

Science as Inquiry Standards:

- Abilities Necessary to Do Scientific Inquiry

  Students observe the simple machines in the Jam Maker.

  Students base explanations on what they observe.

Physical Science Standards:

- Transfer of Energy

  Energy is associated with mechanical motion and can be transferred by simple machines such as levers and gears.

MATERIALS

For the Procedure

Per group of 4 students

- Parker Brothers Grape Escape™ game

This activity can be done in an independent learning center if only one game is available.

Per class
- flathead screwdriver

SAFETY AND DISPOSAL

Games can be put back in their boxes still partially assembled to save time the next time the activity is done. No special safety procedures are required.

GETTING READY

Depending on the age of the students, you may wish to build the Jam Makers prior to student use as the instructions may be difficult for younger students to follow. The first Jam Maker may take 5–10 minutes to assemble. After that, the rest of the Jam Makers can be assembled more quickly. If an extra Jam Maker is available, disassemble the base so students having trouble figuring out the gear system can take a closer look at it. To do this, push down with a screwdriver on the green tabs in the center of each of the two large purple disks and lift the purple part up. The purple disks can be snapped back in place later with no permanent damage to the toy.

INTRODUCING THE ACTIVITY

Show the game box to the students and tell them that they will be analyzing this game to see how it works. Briefly review the different types of simple machines involved in the game: gear, lever, wheel and axle, and inclined plane.

PROCEDURE

Options:

- One option for the Procedure is to give groups the Investigation Sheet (provided) and have them work independently. (If only one game is available, groups can complete their Investigation Sheets at a learning center.) Conduct a whole class discussion of the results when all groups have completed their Investigation Sheets. (If you choose the learning center method, give groups a few minutes to review their Investigation Sheets just prior to the class discussion.)

- An alternative to passing out the Investigation Sheet is to refer to it yourself and issue the instructions one at a time to the whole class. Have groups share results and answer the discussion questions before you proceed to the next step. You may want to ask students multiple questions about what they have done and observed to draw out discussion.

EXPLANATION

The following explanation is intended for the teacher's information. Modify the explanation for students as required.

A gear, one of the simple machines, is a special kind of wheel and axle. A gear is a wheel with teeth. The teeth of one gear mesh with another, allowing both gears to turn. Gears that mesh must turn in opposite directions. The Jam Maker of this game contains five gears as shown in Figure 1. These gears are used to transfer the force applied to the handle. The three outer gears under the handle, saw, and rolling pin turn in the same direction, while the two center gears with which they mesh turn in the opposite direction.

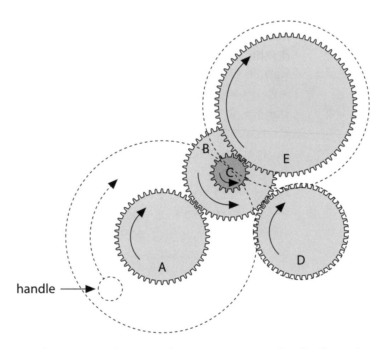

Figure 1: The Jam Maker uses five gears to transfer the force from the handle. When gear (a) moves clockwise, gears (b) and (c) move counterclockwise. These gears then make gears (d) and (e) move clockwise. Note that while (b) and (c) are separate gears, they are part of the same piece.

The size of each gear is also important. The handle must be turned only one complete turn for the saw to move out and back but must be turned four complete turns in order for the rolling pin to move out and back to squish a grape. Thus, we know that all the gears in the transfer chain for the saw are the same size. On the other hand, somewhere in the gear transfer chain for the rolling pin, one gear must be four times larger than the one that is pushing it. A gear that turns four times compared to one that turns once has a circumference one-fourth that of the larger gear. If you count the teeth of both sizes of gears in this transfer chain, you will find that Gear E has 72 teeth and Gear C has 18.

Several other simple machines are also found in the Jam Maker. The scissors, wrench, and rolling pin arms are levers pivoted near the center. The saw is a lever pivoted near one end. The rolling pin is a wheel and axle. The wrench that moves the boot rides up and down an inclined plane.

Simple machines can be examined from both a forces point of view and an energy point of view. Both the gears and levers transfer energy from one part of the Jam Maker to another. This is discussed further below for each "squisher" individually.

 The following part of the Explanation will be much easier to understand if you have the game in front of you while you read.

Scissors

On top of Gear A, which has the handle, is a plate with an off-center circular ridge on it. The edge of the scissors rests against that ridge and moves as the plate moves. A rubber band runs from one blade to the handle of the other blade. When the rubber band is not stretched, the scissors are open. As you turn the handle, the plate exerts a force on one handle of the scissors, moving it away from the center of the circle and stretching the rubber band. Since the other side of the scissors is fixed in position, this motion closes the scissors. As the circular ridge moves away from the scissors, the rubber band pulls the scissors open again. Without the rubber band, the scissors would stay closed. From an energy point of view, you do work as you turn the handle. Part of the work is stored in the rubber band as elastic potential energy, and part goes directly into kinetic energy of the scissors.

Saw

The pink base of the saw rests upon Gear E, which has a knob on it. The saw's base has a slot that the knob fits into. As the gear turns, the knob moves around, and the base and saw move back and forth. The purpose of the rubber band on the saw is to create a force to push the saw down against the grape. As the handle is turned, it turns a gear beneath it. The handle gear meshes with Gear B, which is visible. Gear B meshes with Gear E, beneath the saw, turning that one also. Thus, the force on the handle is transmitted to the base of the saw. The work done as the handle is turned becomes the kinetic energy of the gears and saw.

Rolling Pin

On the green Jam Maker base is a row of teeth. Teeth on the orange rolling pin fit between the teeth on the base. The yellow base fits on a ring on the purple plate covering Gear E, which supplies the motion for the rolling pin. As Gear E turns, the yellow base moves back and forth. The center gear assembly has a small gear mounted on top of a large gear and is rigid so that the two gears must turn together. Rather than mating with the bottom gear (Gear B), as the saw gear does, the rolling pin gear mates with Gear C, the small upper gear. Gear C has only one-fourth as many teeth as all the other gears in the system. Thus, when the handle and Gear C turn once, Gear E, under the rolling pin, goes only one-fourth of the way around. It takes four handle cranks to make the rolling pin go through one complete cycle.

Boot

The boot does not get its energy directly from a gear. The pink wrench that rests on top of the yellow base of the rolling pin is connected to the boot. A rubber band runs from the top of the blue tree to the bottom of the boot where the wrench is attached. As the yellow base of the rolling pin moves outward, one end of the pink wrench is forced up an inclined plane. The other end of the wrench pulls the boot down, stretching the rubber band. The wrench receives energy as it is lifted. That energy is transferred to the rubber band and stored as elastic potential energy. As the free end of the wrench moves back down the inclined plane, the stored energy in the rubber band is transformed to gravitational potential energy as the boot moves back up the tree. (A little kinetic energy is involved in both the up and the down processes, but the amount is small because the boot moves very slowly.) Without the rubber band, the boot would not move upward.

EXTENSION

Take the base of the game apart. Look at the gears. Count the number of teeth in each gear and determine the ratios between the gears. How do they compare? Does this agree with what you found when you counted the number of turns of the handle required to make each section go through a complete cycle?

CROSS-CURRICULAR INTEGRATION

Language arts:

- Imagine that you are a grape in this game. Write a story describing your adventures. How would you escape being squashed?
- Write a skit that explains how gears and other simple machines help us.

Life science:
- Investigate how gears and other simple machines are used in the health profession. Some interesting examples include artificial limbs, traction devices, and various exercise machines used in physical therapy.

Music:
- Write a song to the tune of "Heard it Through the Grapevine" that describes the simple machines used in Grape Escape.

FURTHER READING

Kirkpatrick, L.D.; Wheeler, G.F. *Physics: A World View*, 2nd ed.; Saunders: Philadelphia, PA, 1995. (Teachers)

Zubrowski, B. *Wheels at Work: Building and Experimenting with Models of Machines;* Beech Tree: New York, NY, 1986. (Students)

CONTRIBUTORS

Cris Cornelson, Miami University undergraduate student, Oxford, OH; Teaching Science with TOYS, 1992–93.

Dan Masterman, Los Angeles Unified School District, Los Angeles, CA; Teaching Science with TOYS, 1996.

Richard Boyansky, Camden Central School District, Camden, NY; Teaching Science with TOYS, 1996.

HANDOUT MASTER

A master for the following handout is provided:
- Investigation Sheet

Copy as needed for classroom use.

Squish 'em, Squash 'em, Squoosh 'em
Investigation Sheet

Assemble the Grape Escape game according to the directions. Make several grapes to use during your exploration of gears and energy.

1. Start at one of the "squisher" locations. Slowly turn the handle of the Jam Maker and observe what happens. Repeat this at each location. In the spaces below, explain what is happening to the grape at each "squisher." You may use words, drawings, or both.

Scissors	Saw

Rolling Pin	Boot

2. Identify any simple machines you see.

_____

_____

_____

3. The scissors have a rubber band that runs from one scissor blade to the handle of the other scissor blade. What is the function of this rubber band?

_____

_____

_____

   How would the scissor squisher be different without the rubber band?

_____

_____

_____

4. Where in the Jam Maker is elastic potential energy stored?

_____

_____

5. At the rolling pin station, the green Jam Maker base contains a row of teeth. Teeth on the orange rolling pin mesh with the teeth on the green base. How does this help squish the grape?

_____

_____

6. Slowly turn the handle of the Jam Maker and count the number of turns it takes to complete each "squish" action.

   Scissors _____

   Saw _____

   Rolling Pin _____

   Boot _____

7. Draw the layout of the gears beneath the Jam Maker. You may not be able to see all the gears, but you should be able to deduce where they are. Are all the gears the same size?

   _____

8. Turn the handle to determine in which direction the gears move. Add arrows to the drawing above to show the direction the gears are moving. Are all the gears moving in the same direction?

   _____

9. The saw uses a system of three gears meshed together to squish the grape. Describe how energy is transferred through the system to squish the grape.

   _____

   _____

   _____

   _____

   _____

10. Is a gear directly connected to the boot? _____ What causes the boot to squish the grape?

    _____

    _____

    As with the scissors, a rubber band is also attached to the boot. What does this rubber band do?

    _____

    _____

    _____

Activities Indexed by Science Topics

National Science Education Standards Matrix

	Activities		

This matrix shows how the activities in this book relate to the National Science Education Standards. The standards are taken from *National Science Education Standards;* National Research Council; National Academy: Washington, D.C., 1996.

	What Makes It Go?	The Toy That Returns	How Much Energy?
Science as Inquiry Standards			
Abilities Necessary to Do Scientific Inquiry			
Identify questions that can be answered through scientific investigations.			
Design and conduct a scientific investigation.	✔	✔	
Use appropriate tools and techniques to gather, analyze, and interpret data.			✔
Develop descriptions, explanations, predictions, and models using evidence.	✔	✔	
Think critically and logically to make the relationships between evidence and explanations.			✔
Recognize and analyze alternative explanations and predictions.			
Communicate scientific procedures and explanations.		✔	
Use mathematics in all aspects of scientific inquiry.		***	
Physical Science Standards			
Transfer of Energy			
Energy is a property of many substances and is associated with heat, light, electricity, mechanical motion, sound, nuclei, and the nature of a chemical. Energy is transferred in many ways.	✔	✔	✔
Heat moves in predictable ways, flowing from warmer objects to cooler ones, until both reach the same temperature.			
Electrical circuits provide a means of transferring electrical energy when heat, light, sound, and chemical changes are produced.			
In most chemical and nuclear reactions, energy is transferred into or out of a system. Heat, light, mechanical motion, or electricity might all be involved in such transfers.			
Motions and Forces			
The motion of an object can be described by its position, direction of motion, and speed. That motion can be measured and represented on a graph.			
An object that is not being subjected to a force will continue to move at a constant speed and in a straight line.			
If more than one force acts on an object along a straight line, then the forces will reinforce or cancel one another, depending on their direction and magnitude. Unbalanced forces will cause changes in the speed or direction of an object's motion.			

	Exploring Energy with an Explorer Gun	Pop Can Speedster	Ladybug, Ladybug, Roll Away	Rubber Band Airplane	Slingshot Physics	The Catapult Gun	Loop-the-Loop Challenge	Homemade Roller Coaster	Bounceability	The Energy Transformation Game	Drop 'n' Popper	Apply Your Energy Knowledge	Doc Shock	Make Your Own Motor	Chemical Energy Transformations	Simple Machines with LEGO	Get It in Gear with a LEGO Vehicle	Squish 'em, Squash 'em, Squoosh 'em
				✔									✔					
	✔	✔		✔	✔	✔	✔	✔	✔		✔	✔	✔	✔	✔		✔	✔
	✔			✔	✔	✔		✔	✔									
	✔	✔	✔			✔	✔	✔			✔	✔	✔				✔	✔	✔
		✔			✔													
		✔						✔	✔	✔		✔				✔		

*** This standard is addressed by the energy unit as a whole, not by individual activities. ***

	Exploring Energy with an Explorer Gun	Pop Can Speedster	Ladybug, Ladybug, Roll Away	Rubber Band Airplane	Slingshot Physics	The Catapult Gun	Loop-the-Loop Challenge	Homemade Roller Coaster	Bounceability	The Energy Transformation Game	Drop 'n' Popper	Apply Your Energy Knowledge	Doc Shock	Make Your Own Motor	Chemical Energy Transformations	Simple Machines with LEGO	Get It in Gear with a LEGO Vehicle	Squish 'em, Squash 'em, Squoosh 'em
	✔	✔	✔	✔	✔	✔	✔	✔	✔	✔	✔	✔				✔	✔	✔
													✔	✔				
														✔	✔			
				✔	✔		✔				✔							

About The Author

Terrific Science Press is a nonprofit publisher housed in the federally and state-funded Center for Chemistry Education (Miami University Middletown in Ohio). At the Center, educators and scientists have worked together since the mid-1980s to provide professional development for teachers through innovative approaches to teaching hands-on, minds-on science.